Modems in Data Communications

P R D Scott

PUBLISHED BY NCC PUBLICATIONS

British Library Cataloguing in Publication Data

Scott, P R D
Modems in data communications.
1. Modems
I. Title
001.6'443 TK5105

ISBN 0-85012-243-0

Published in 1980 by:

NCC Publications, The National Computing Centre Limited,
Oxford Road, Manchester M1 7ED.

Typeset in 10pt Press Roman by Focal Design Studios Limited,
New Mills, Stockport, Cheshire.

Printed by Hobbs the Printers of Southampton (1103/87)

ISBN 0-85012-243-0

Foreword

Someone said "if you think education is expensive try ignorance". In the field of modems and data communications that aspect of education which requires the availability of good books has been inadequately served; I believe this book will help to restore the balance.

There can be few activities within the broad business sector of electronics which are growing more rapidly and which can be expected to become so pervasive as those made possible by data communications: viewdata, electronic funds transfer, the office of the future, home computers, distributed data processing and so on. Five years ago there were those who were predicting that the modem would be dead by 1980 and that communication would be digital all the way. It is reassuring for those of us in the industry to note that the NCC believes that the modem has enough of a future to justify publishing a book largely devoted to it.

Many of the existing books on data communications were written by researchers or academics and were dedicated to the spread of information on its theoretical aspects, preparing the ground as it were for new theories and ideas. The user of data communications however, the communication or data processing manager, is much more interested in what data communication can achieve and how it should be used, right down to the apparently humble questions of what happens at which pin of a connection on the back of a modem. This practical information needs sufficient support on basic theory and the telephone network itself to provide a well-rounded whole. I feel sure that this book provides just such a balanced mixture.

Barry Stuttard
Technical Director, Racal Data Communications Group
April 1980

Acknowledgements

I am grateful to the following who gave their time, and allowed me to benefit from their practical experience of building and running data communications systems:

Rod Bird	Lucas Group
Martin Jones	Lucas Group
Jeff Edwards	British Steel
V Rajeswaren	Reuters
Martin Renard	Compower
Roger Paul	BL Systems
Jean McGregor	North Thames Gas
R Banks F Ball	Post Office Telecommunications Headquarters

I would also like to thank Case, Racal Milgo, the Post Office and those individuals who read an early draft of this book and provided valuable comment. This does not, however, absolve me from taking responsibility for all errors and omissions.

The following modem suppliers were good enough to make available operating manuals and technical data:

Computer and Systems Engineering Ltd
Cole Electronics Ltd
IAL
IBM (UK) Ltd
Nolton Communications Ltd
Racal Milgo Ltd

P R D Scott

The Centre acknowledges with thanks the support provided by the Computers, Systems and Electronics Requirements Board (CSERB) for the project from which this publication derives.

Preface

This book stemmed from a project to investigate the degree to which modem standards are implemented in practice. There are some two dozen standards of various types relating to the terminal/modem interface, ranging from international (ISO) standards and CCITT recommendations to trade standards and manufacturers' individual standards, many of which contain implementation options which could be expected to cause problems. Yet, in practice, the terminal/modem interface seems to give rise to very few difficulties; users find it one of the least troublesome areas within a data communications system. It is generally the case that any approved terminal will interwork with any modem. Such compatibility is rare in computing, and it is interesting to see why it occurs in this instance. There are three main reasons.

1 The standards are good

They are good in the sense that the connection between the terminal and the modem is a sensible interface to standardise. It is the boundary between the computing system and the data transmission system, and is thus a 'natural' boundary. The standards are also good in that they are mature; they have been around since the 1960s and by now have had most of the bugs removed. Also, this maturity or stability has provided the modem manufacturers with a sure market, which in turn has given them the confidence to invest in the development of new techniques – such as the incorporation of microprocessors.

2 The modem manufacturers have been accommodating

The various options available in the different standards could so easily have led to incompatible terminal/modem combinations, but the modem manufacturers, by providing strapping or switching facilities to allow the modem to be configured in a variety of ways, have made it possible for their modems to accommodate most data terminal configurations.

3 The PTTs and common carriers have acted as certification authorities for the standards

This is the single most important reason why modem standards are so well implemented in practice. The PTTs usually insist on the use of their own modems on the public switched telephone network (PSTN), and therefore also require that terminals connected to these modems are of an approved type and conform to the international terminal/modem interface standards. The existence of a large terminal population with a PTT-approved standard interface ensures in turn that private modem suppliers also conform to the standards, even though in their case PTT approval may be confined to the transmission line interface only.

In the UK, this certification role of the Post Office is often criticised; complaints are usually about the cost, the delay, or the fact that certification is mandatory and that the Post Office has the power to deny equipment suppliers permission to use national transmission facilities. However, one has only to look at the level of standards observance in any other area of computing to see how effective this certification is in permitting successful equipment interconnection. It is an open question whether the same results would be obtained from a different certification scheme – but this book does not intend to go into such issues.

SCOPE

The book is aimed at the modem user. It describes what a modem does, and how it interacts with other components in a data communication system, but it does not attempt to describe the inner workings of a modem.

It can be thought of in four main sections. The first section (Chapters 1 and 2) deals with the basics of data transmission, and describes the environment in which the modem has to operate – the telephone network. The second (Chapters 3 and 4) deals with modulation theory, and the various techniques that have been standardised internationally. The third (Chapters 5, 6 and 7) deals with all aspects of the terminal/modem interface – protocols, connectors and electrical characteristics. The fourth (Chapters 8 and 9) describes different types of modem and ancillary equipment, and data networks.

The book is not a buyer's guide to modems, but it will help the potential user to understand some of the jargon in the modem manufacturers' advertising material. Another NCC book *Selection of Data Communications Equipment* (Nichols and Jocelyn 1979) deals with the criteria on which the choice of modems should be based. NCC also publishes the Computer Hardware Record, a series of bulletins on different types of computer equipment. The issue on modems and acoustic couplers lists all the modems available in the UK, together with supplier details and a brief specification. The Computer Hardware Record is available on subscription, and further details should be sought from the NCC Information Service.

Contents

1 Data Transmission Theory

INTRODUCTION

The first transatlantic telegraph cable was successfully laid in 1858. Such was the impact of this breakthrough in communications – the beginning of international data transmission – that it merited the attention of Queen Victoria, who sent greetings to the then President of the United States, President Buchanan. The 100-word message is reported to have taken 16 hours to complete – not a very impressive data rate even by the standards of the time, when 25 words/minute was considered normal on land.

What the early pioneers found was that when a voltage was applied to one end of the cable, it did not appear instantaneously at the other end but instead built up gradually, only reaching a steady state value after a period of time. When the voltage was removed, the receiving-end voltage did not drop suddenly, but decayed slowly. The cable was acting like a sponge, storing electricity when a voltage was applied and allowing it to leak away when the voltage was removed. This is the property we now call capacitance. It made for extremely slow signalling; each signal had to be maintained long enough for it to reach a detectable level and then had to be given time to die away before a second signal could be sent.

In an attempt to overcome the 'sluggishness' of the cable, and to achieve a sensible data transmission rate with the relay-operated receivers then employed, higher and higher voltages were applied until eventually, 11 weeks after installation, the insulation of the cable broke down and the cable was rendered useless. It was 8 years before another cable was laid across the Atlantic. The interval proved to be of value, however, in that it gave time for the theory of cable transmission to be consolidated and disseminated. One of the leading minds to be applied to the problems of telegraphy was that of William Thomson, later Lord Kelvin, and he in turn was influenced by the theories of a Frenchman, called Fourier, on heat transfer. As we shall see, Fourier's work had wider application than perhaps he realised, and is today fundamental to the theory of data transmission.

In 1866, telegraph transmission across the Atlantic was resumed, and for the next 90 years the only transatlantic submarine cables laid were telegraph cables. The first telephone cable, TAT-1, was not put down until 1956.

On land, developments took a different course. Following Bell's invention of the telephone in 1876, telephone networks expanded rapidly and soon overtook the telegraph networks both in terms of size and traffic carried, despite the pessimism of the telegraph operators who doubted whether the public could be trusted to handle 'technical communications equipment'.

Demand for data transmission as we know it today began in the 1950s, with the growth of computing. Low-speed telegraph circuits did not satisfy the needs for long, and means were soon sought to exploit the higher information-carrying capacity of telephone circuits. Digital data is inherently different from analogue speech signals, and to allow data to be transmitted down a telephone circuit a conversion device or 'modem' was developed. Just how a modem behaves is the subject of this book, but before that there are some basic facts about digital data which need to be covered.

THEORY OF DATA TRANSMISSION

The digital data signal generated by a typical data terminal is a square wave like the one depicted in Figure 1.1. The two states could be two different currents or two different voltages. One legacy of early telegraphy practice is the use of the terms 'mark' and 'space' for the 1 and 0 levels respectively, from the days when signals were indicated by an ink mark, or lack of it, on a strip of paper. A two-state signal is known as a binary signal.

Figure 1.1 actually shows a *polar* signal, since the two states have equal positive and negative values. Since time increases towards the right, the bits in Figure 1.1 would be transmitted in the order 0, 1, 1, 0, 1, 0, 0. By convention in data transmission, the low-order bit is always transmitted first.

The greatest rate of change of information in a binary signal occurs when alternate 0s and 1s are transmitted. If the bit rate is N bit/s, it can be seen from Figure 1.2 that the binary signal 0 1 0 1 0 1 . . . carries information at the same rate as a sine wave of frequency N/2 Hertz.* However, if we were to transmit this data as a square wave through a channel with an upper frequency limit of N/2 Hz, we would find that the square wave was considerably rounded off. If we look at the wave forms in Figure 1.2c we can begin to see why. The time taken for the signal to rise from its minimum to its maximum value is shown as 'r'. The higher the frequency of the wave, the shorter 'r' is. Now a square wave changes state almost instantaneously (ie r is small), and this implies the presence of very high frequencies. In fact a full analysis of a cont-

* Hertz, abbreviated Hz, a unit of frequency. 1 Hz = 1 cycle per second.

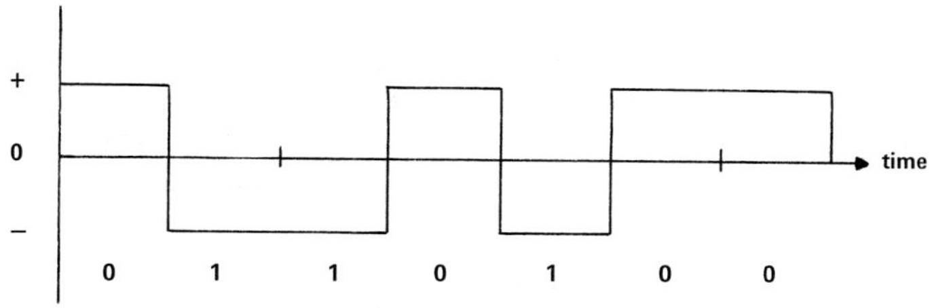

Figure 1.1 Polar Binary Signal

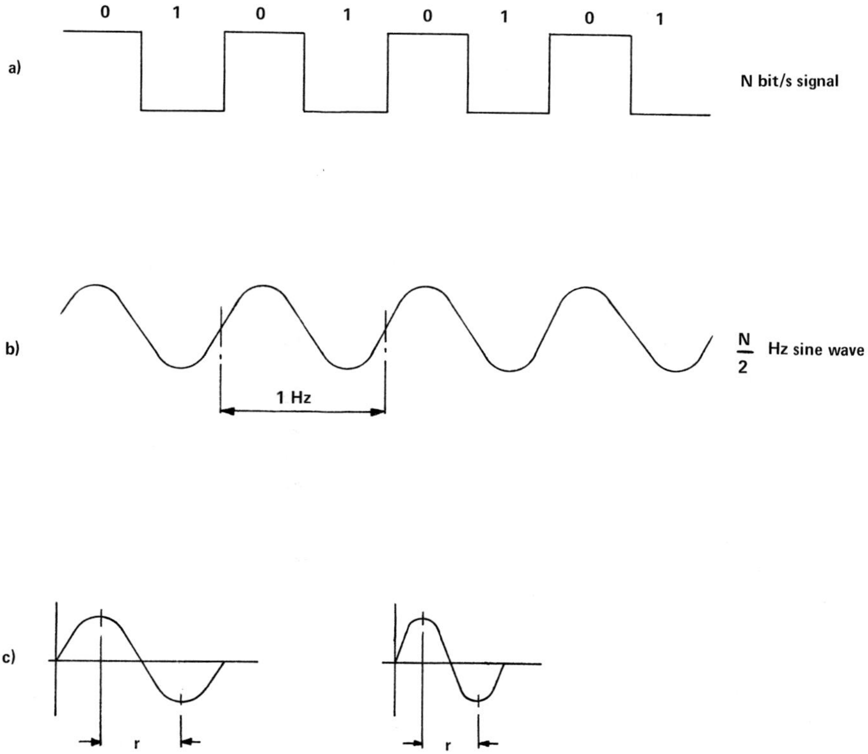

Figure 1.2 a) Square Wave, Alternate 0s and 1s b) Sine Wave Equivalent
c) Risetime of Waves of Different Frequency

inuous square wave reveals that it is composed of a whole series of harmonically related sine and cosine waves of differing amplitudes. Such a series is known as a Fourier series. Thus, to transmit a true square wave, that is one with vertical edges, a channel of infinite bandwidth would be required. ('Bandwidth' refers to the band of frequencies that a channel can carry; it is measured in Hertz.) We can now appreciate that the observations made on that first transatlantic telegraph cable are consistent with a very low bandwidth circuit.

Fourier also dealt with the analysis of a *single* square pulse. In this case the component frequencies are not a series of discrete frequencies, but a continuous spectrum. The amplitude of each component frequency is given by the Fourier Transform of the pulse. The Fourier Transform of a single square pulse is a curve of the form (sin x)/x, as shown in Figure 1.3. This curve gives the amplitude of all the component frequencies of the pulse. Note that for a pulse of duration T, certain frequencies, at f, 2f, 3f, etc, have zero magnitude (f = 1/T).

Suppose, however, we had started not with a square pulse, but with one of (sin x)/x shape (Figure 1.4a). What would be the component frequencies of that pulse? It turns out that the Fourier Transform of a (sin x)/x pulse has a spectrum with a sharp cut-off at one particular frequency, as shown in Figure 1.4b. Since our transmission channels always have a finite bandwidth, this would seem to be the ideal pulse to use to avoid distortion. Nyquist showed in 1928 that the maximum repetition rate of such pulses over a perfect channel was 2f pulses per second, where f is the bandwidth of the circuit in Hertz. The period 1/2f, which is the time between pulses, is known as the Nyquist interval. The signalling rate of 2f pulses per second is known as the Nyquist rate.

Although pulses of (sin x)/x shape are ideal in theory, there are other shapes which are more tolerant of the deficiencies of practical transmission systems. One of these is based on the spectrum of what is known as a raised cosine pulse. A raised cosine pulse is sketched in Figure 1.5a, together with its spectrum (Fourier Transform) in Figure 1.5b. A pulse based on the spectrum of the raised cosine pulse (Figure 1.5c) has a Fourier Transform as shown in Figure 1.5d. It can be seen that the penalty paid in using a pulse of this shape is that twice the bandwidth is needed compared to the (sin x)/x pulse shape. However, one of the important properties of this pulse shape is that the level is only above the halfway point for half the pulse duration, and pulses of duration T can be sent at intervals of T/2 seconds. Figure 1.6 shows how the bit pattern 0 1 1 0 1 0 0 would look in raised cosine spectrum form, together with the sampling instants.

These more sophisticated pulse shapes are not seen at the output of a data terminal, but may be generated by the modem or other transmission equip-

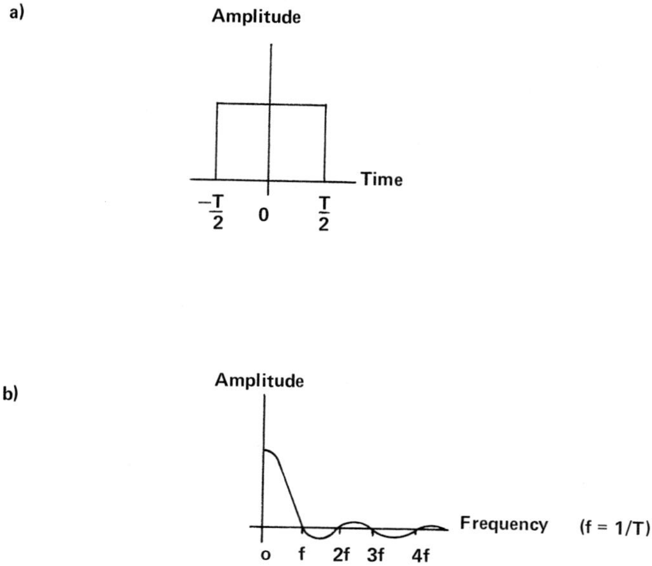

Figure 1.3 Single Square Pulse (a), and its Fourier Transform (b)

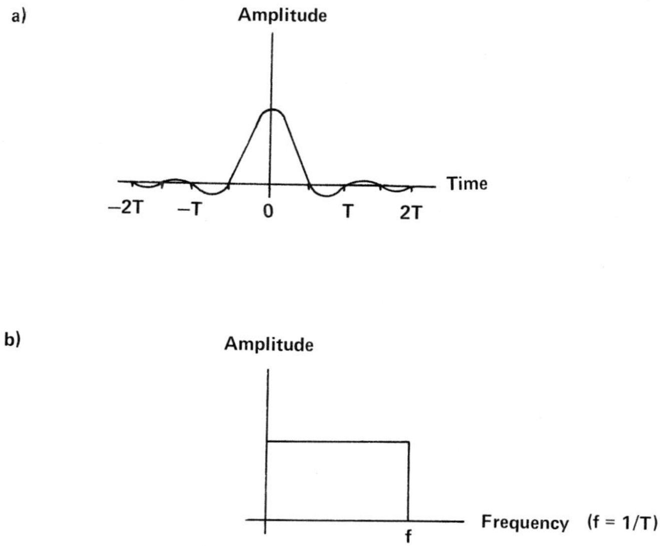

Figure 1.4 Sinx/x Pulse (a), and its Fourier Transform (b)

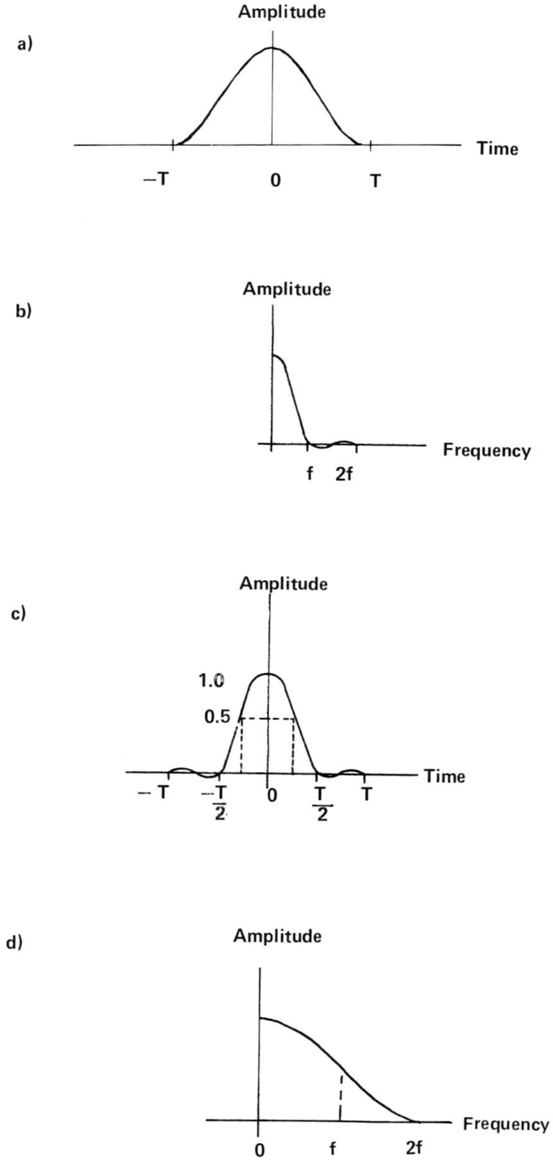

Figure 1.5 a) Sketch of Raised Cosine Pulse, and b) its Fourier Transform
c) Pulse based on FT of Raised Cosine Pulse (b), and its Fourier Transform (d)

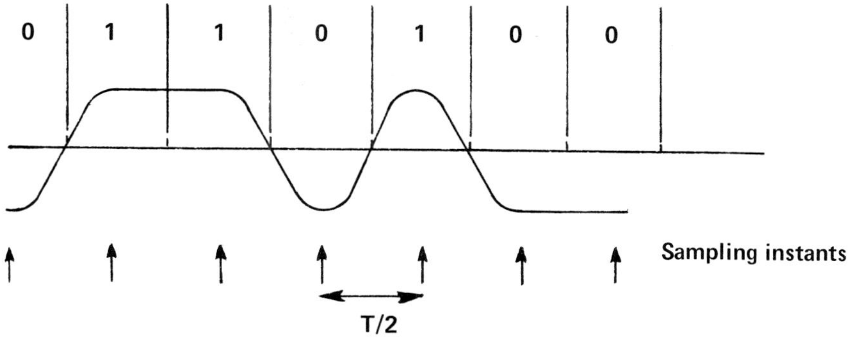

Figure 1.6 Binary Waveform using a Pulse Shape based on the FT of a Raised Cosine Pulse

ment to which the terminal is connected.

Synchronous and Asynchronous Transmission

The data generated by a terminal is normally character-oriented and is output in serial form as a series of equal duration bits. Characters may be output either at random intervals, when the transmission method is called start-stop or asynchronous, or at predetermined times when the method is called synchronous (or more correctly) isochronous. In both cases, the signal is known as a *baseband* signal, because it contains frequencies down to dc.

Start-Stop Transmission

In the 'idle' state, a start-stop terminal maintains a MARK (binary 1) condition on the line. Any character transmitted is preceded by a SPACE (binary 0) condition of 1 bit duration, to indicate the start of the character. Detection of this 'start' bit by the receiver initiates a timing mechanism which indicates when the incoming line signal is to be sampled (Figure 1.7a). Clearly the receiver timing mechanism must be set to the same nominal signalling rate as the transmitter, and must know the number of bits in the character.

Figures 1.7b & c show the effect of a difference in timing between transmitter and receiver. If the receiver timing is too fast, it will sample one bit twice, and if it is too slow it will miss one bit altogether. For correct decoding the receiver needs to be accurately synchronised to the incoming signal for

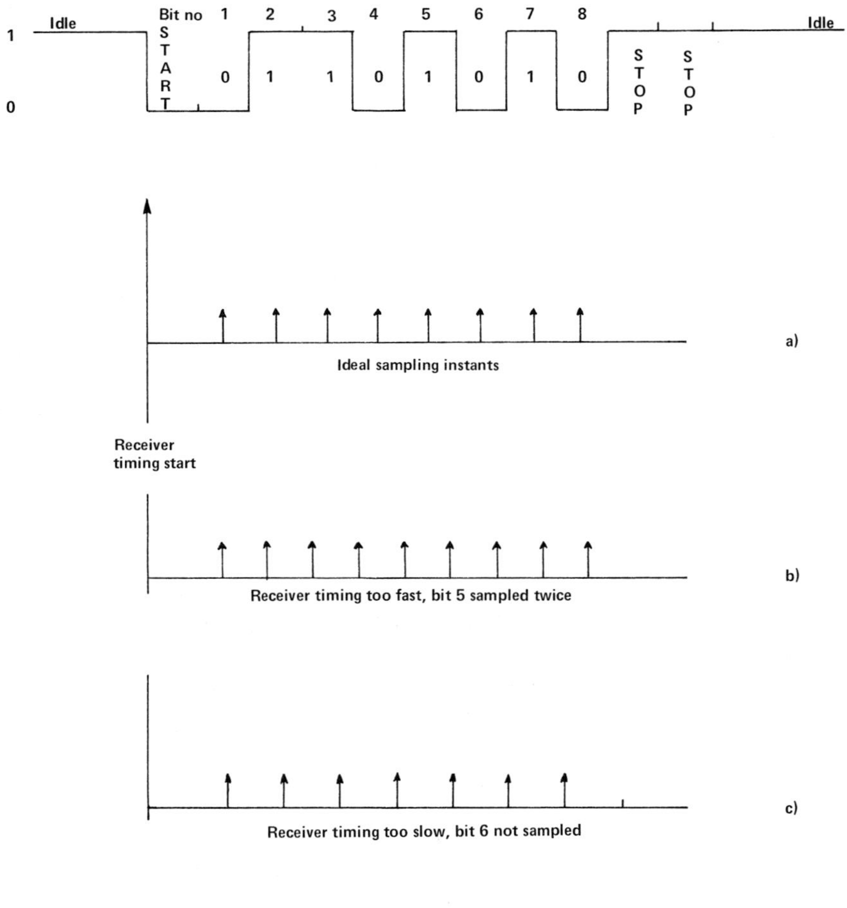

Figure 1.7 Start-stop character structure,with
a) Correct receiver timing
b) Receiver timing fast
c) Receiver timing slow

the duration of a character. The stability of the timing mechanism governs the maximum character length.

The common teleprinter code uses 5 data bits, with one start bit and 1½ stop bits. The stop bits at the end of a character allow the timing mechanism to reset itself ready for the next character. Modern data terminals use 8 or 9 bit codes, with 1, 1½ or sometimes 2 stop bits.

Start-stop transmission is used widely by terminals operating at the lower data rates, that is up to and including 1200 bit/s. It is sometimes known as anisochronous operation (although, strictly speaking, anisochronous describes data signals such as facsimile signals where each signal element can have a different length).

Synchronous Operation

For high-speed terminals, the start and stop bits of asynchronous operation present an unwelcome overhead, reducing the throughput of the channel. In synchronous operation, characters are transmitted one after the other in a continuous sequence without any intervening start or stop bits. This is called isochronous transmission. If there is no data to send, special idle characters are transmitted instead. (As an alternative, an isochronous bit stream can be used to carry start-stop data. Davies and Barber* propose the term 'isochronous start-stop' for this mode of operation.)

The clocks in synchronous terminals need to be accurate since they have to remain synchronised over many characters, not just a single one. To maintain synchronism, the receiver clock monitors the transitions in the incoming data and corrects itself by adding or deleting a sampling interval as necessary. To ensure that there will be transitions in the incoming data, it is common for the transmitted data to include special synchronising characters at regular intervals in the bit stream.

On top of synchronisation at the bit level, the receiver also needs to know where in the bit stream a character starts. (Characters are all the same length, so once the start of the first one is known, the position of the rest can be deduced.) Character synchronisation is achieved by sending a group of synchronising characters at the start of each block of data characters sent during the transmission.

The definition of synchronising sequences, block sizes, formatting characters and the like, constitute a protocol. A protocol is in effect an agreed procedure which ensures the orderly flow of data. Intercommunication between synchronous terminals requires conformity to a common protocol. The logic needed to implement this protocol, together with other circuitry, such as a

* D W Davies and D L A Barber. *Communication Networks for Computers*, Wiley 1973.

buffer store for transmitted data, makes a synchronous terminal more costly than an equivalent asynchronous terminal.

Bits and Bauds

We have seen already that a perfect transmission channel of bandwidth B Hz can transmit 2B signals per second. The signalling rate, or modulation rate, is measured in bauds; a baud is one signal per second. If we are using a binary signal as in Figure 1.2, the bit rate is the same as the baud rate.

Suppose, however, that our original binary data had been coded in another way. We could have used four different signalling levels – say ± 2 and ± 1 units – instead of just the two levels used previously. The binary data could have been coded by splitting it up into pairs of bits, and assigning one level to each possible bit combination in the pair, as in Table 1/1.

Bit combination	Level
00	+2
01	+1
10	−1
11	−2

Table 1/1

Then a waveform like the one in Figure 1.8 would represent the data 0 0 0 1 1 0 1 1 0 0. Although the signalling rate is the same as before, the data rate has been doubled since we can now carry 2 bits per baud. Thus, in a 4-level system, the bit rate is twice the baud rate. In general, we can say that:

Bit rate = baud rate x number of bits per baud

To code 1 bit per baud we need a 2-level signal, to represent 0 and 1; to code 2 bits per baud we need a 4-level signal, to represent 00, 01, 11, 10; and so on. Expressed mathematically, we need 2^n levels to code n bits or, putting this the other way round, a system with m levels can code $\log_2 m$ bits ($m = 2^n$).

Since a channel of bandwidth B can carry 2B signals/second, the capacity (maximum bit rate) of a perfect m-level transmission channel of bandwidth B is given by:

$$
\begin{aligned}
C &= \text{baud rate x no of bits/baud} \\
&= 2B \log_2 m \text{ bit/s}
\end{aligned}
$$

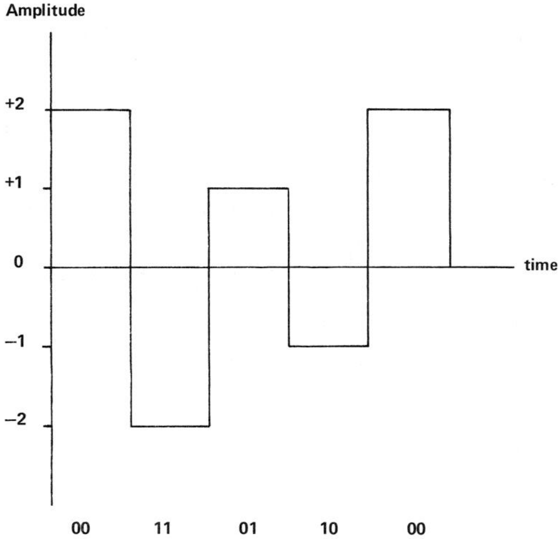

Figure 1.8 4-Level Data

This equation is of theoretical interest only, because no channel is perfect. A real channel exhibits noise and Shannon has shown that the capacity of a noisy channel is given by:

$$C = B \log_2 (1 + S/N)$$

where S/N is the signal to noise power ratio.

For a typical telephone circuit with B = 3000 Hz, S/N = 1000:1 (30 dB*)

$$C = 3000 \times \log_2 (1 + 1000)$$
$$\simeq 30000 \text{ bit/s}$$

Even this equation gives a capacity much greater than can be achieved in practice, because it does not take into account all the other impairments encountered in real transmission systems.

* dB – abbreviation for deciBel. A decibel represents a power ratio, and is defined as $10 \log_{10} (P_1/P_2)$ where P_1 and P_2 are the two power levels being compared. (If P_2 = 1 milliwatt, the abbreviation becomes dBm.)

Example If the signal power is 1 milliwatt and the noise power is 1 microwatt, the signal to noise ratio is 1000:1 or $10 \log_{10} \frac{1000}{1} = 30$ dB. Alternatively, the noise power could be expressed as –30 dBm, and the signal power as 0 dBm.

2 UK Data Transmission Services

UK TELECOMMUNICATIONS NETWORKS

Telecommunications in the UK, as in all other countries of the world, is governed by the demands of the telephone service. The one vestige of the telegraph systems from which it all began is the telex network which, although functionally separate from the telephone service in the UK, nevertheless shares transmission plant with it. Until recently, telex was the one and only *digital* transmission system, but its speed and code limitations mitigated against its widespread use for data communications. Instead requirements for data transmission facilities had to be met by 'stretching' facilities provided for telephony. The telephone network has proved to be resilient enough to permit such stretching, and has coped reasonably well with the demands for data transmission made upon it.

Today most developed countries have or plan to have dedicated data networks using digital transmission and switching, and many see this as a first step towards an integrated digital network providing the whole range of telecommunications facilities, including speech, data, facsimile and more. It is salutory to realise, however, that there are 160 telephone connections for every one data connection in the UK, and that telephony traffic is likely to remain the major revenue-earner for the PTTs and common carriers for at least the next two decades. It will be several years before the data traffic carried on the new data networks exceeds the data traffic carried on the telephony network. Modems – the devices which provide the means for transmitting data over analogue telephony circuits – may become superfluous in the long term, but have an important role to play in the meantime.

In describing the telecommunications networks in the UK, the emphasis of this chapter is on the telephone network because that is the network over which modems operate, and it is important to have a grasp of its structure and organisation.

TELEX AND PSS

Telex and PSS are both digital data services – albeit at two extremes. Telex is a very basic inflexible low-speed service; while PSS is a sophisticated high-speed service. Being digital however, neither service uses modems (except for access to the service in the case of PSS), and so apart from a discussion of interfaces in Chapter 9, these two services are not given further consideration.

THE UK TELEPHONE NETWORK

Topology

The public switched telephone network (PSTN) is a distributed switching network, with the various network nodes (exchanges) arranged in a hierarchy of levels.

At the lowest level is the local exchange. Subscribers – who in most cases are within 3 miles of their local exchange – are connected to it by a pair of wires. Local exchanges vary in size from under 200 lines in remote rural areas, to 20 or 30 thousand lines in large cities. The local exchange acts as an intelligent concentrator; it switches traffic between its own local subscribers and concentrates other traffic onto a relatively small number of routes to either the main trunk network switching centre or to other local exchanges. The degree of interconnection of local exchanges varies, depending on local traffic patterns. Where the traffic between two exchanges is high a direct route is provided; otherwise traffic is routed via an intermediate exchange. In the large cities, dedicated 'tandem' exchanges exist to perform this inter-exchange switching function.

Most towns have a main network switching centre (MNSC), usually co-located with a local exchange. MNSCs are also known as group switching centres (GSCs); they represent the second level in the hierarchy, and there are about 360 of them in the UK, each one serving a group of local exchanges.

Circuits between local exchanges, and between local exchanges and the MNSC, are called junctions. MNSCs are linked together via the trunk network, an interconnected mesh of high-capacity cable and radio links carrying multiplexed circuits. (Multiplexing allows a bearer (eg a cable) to carry more than one channel at a time.) Multiplexing equipment for the trunk network is housed in repeater stations. The term toll circuit, used in the USA, covers both trunk and junction circuits.

Direct point-to-point routes between MNSCs are provided where the traffic warrants it, but otherwise traffic is routed via one or more intermediate exchanges. Successive routeing over several MNSC links degrades the transmission quality on a call, and to avoid this certain routes are provided via a high-level network called the transit network, a fully-interconnected mesh of exchanges linking MNSCs.

At the highest level of the hierarchy are the international exchanges in Glasgow and London, which carry all traffic between the UK and the rest of the world. Figure 2.1 shows the hierarchy of exchanges in diagrammatic form.

Characteristics of Speech Circuits

The human voice contains frequencies up to about 8 kHz. Very low frequencies are generally absent, most of the information being contained in the 100-400 Hz range. The higher frequencies serve to give 'character' to a voice. The provision of bandwidth in a telecommunications network is expensive, and a compromise has to be reached between quality of speech, which calls for high bandwidth, and cost, which demands low bandwidth.

The nominal bandwidth of a speech circuit, ie the band of frequencies it will transmit satisfactorily, is 3.1 kHz, and extends from 300-3400 Hz. On some international circuits the upper limit is reduced to 3 kHz, giving a bandwidth of 2,700 Hz.

Transmission Plant

Local Network

The twisted wire pair used in the local network to connect a subscriber to his local exchange is used for both directions of transmission; it is a '2-wire' circuit. The wire normally used has 0.63 mm diameter conductors, but thicker wire may be used to meet the PO's design requirement of a maximum DC loop resistance of 1000 ohms.

This resistance is frequency-independent, but the wire also has capacitance which causes the total losses to increase with frequency in the manner shown in Figure 2.2. There is no definite limit to the frequencies that can be transmitted; the bandwidth (the range of frequencies carried) is determined by the losses that can be tolerated. The PO design limit for telephone circuits is a maximum loss of 10 dB for the 'local end' (the subscriber to local exchange link).

The DC path is broken at the local exchange by a transmission bridge, which provides AC continuity for audio frequencies but isolates the exchange's power supply from other exchanges.

Junction Circuits

Junction circuits are commonly 2-wire twisted pairs, but unlike local circuits they are usually 'loaded' and often amplified. 'Loading' is the term used when inductance is added to a wire to offset the effect of capacitance and to improve its transmission performance for speech. The added inductance takes the form of loading coils – small coils of wire connected to the cable at

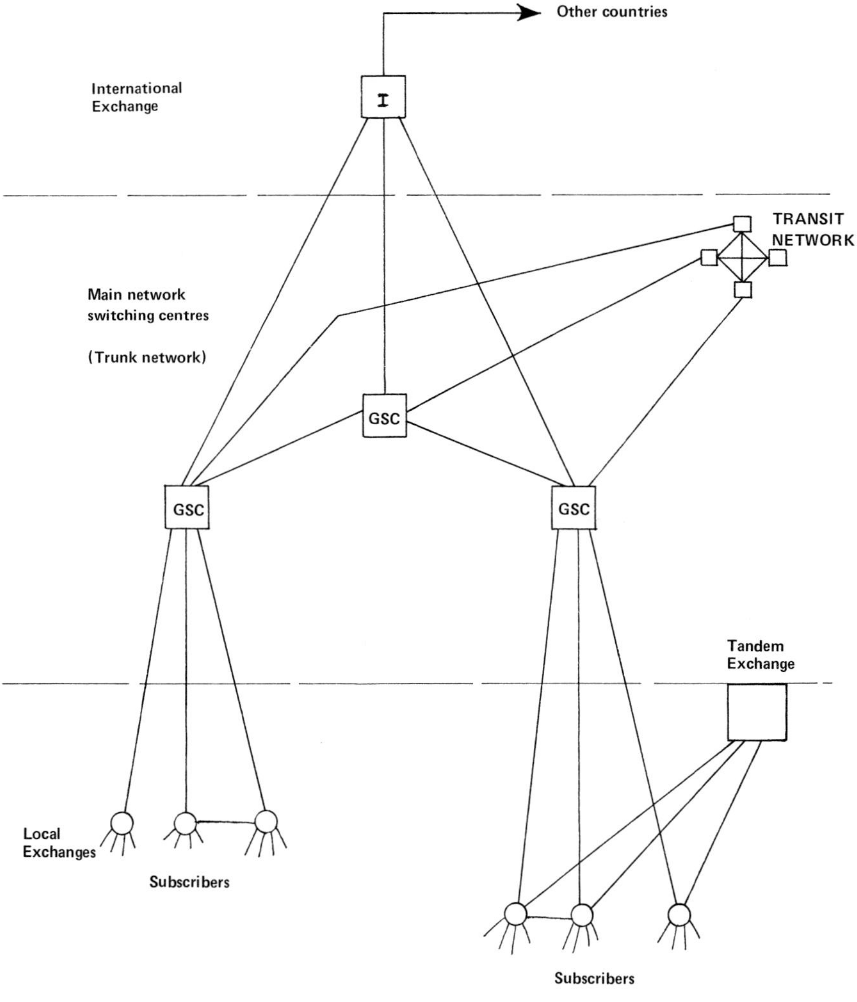

Figure 2.1 Telephone Network Hierarchy

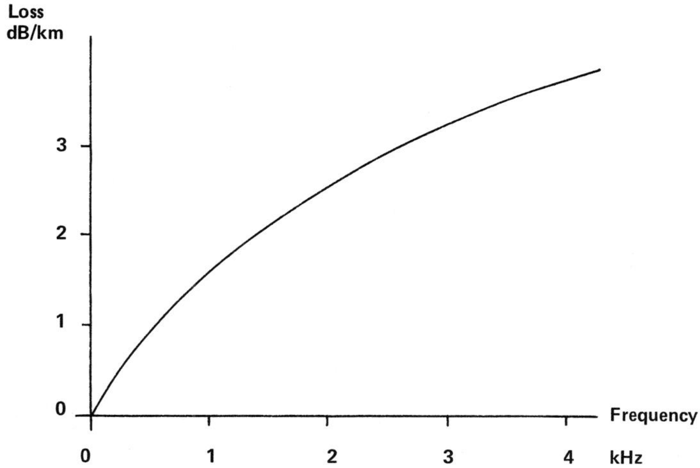

Figure 2.2 Characteristic of Non-loaded Cable

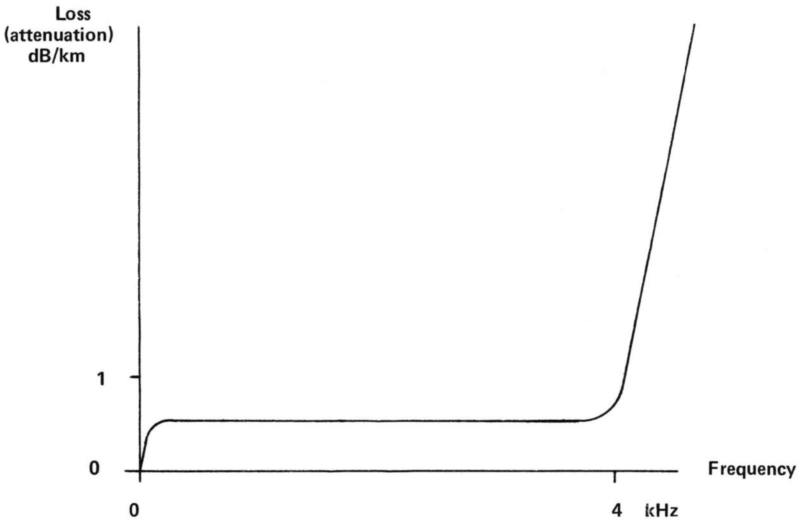

Figure 2.3 Characteristic of Loaded Cable

regular intervals. Figure 2.3 shows the frequency response of a loaded cable; instead of the gradual rise in attenuation exhibited by the unloaded cable (Figure 2.2), the response of a loaded cable is substantially flat up to a certain cut-off frequency after which the loss rises dramatically. A loaded cable thus behaves like a low-pass filter.

2-wire amplifiers allow junction circuits to be extended to about 24 miles using standard cable. For longer junctions, 4-wire circuits are employed because of the technical limitations of 2-wire amplifiers. 4-wire circuits provide separate 'go' and 'return' channels, with separate amplification for each direction.

Increasingly now, time-division multiplexing techniques are finding their way into the junction network, to improve cable utilisation. Pulse code modulation (PCM) is used, the standard system providing 30 speech channels, and operating at an aggregate data rate of 2.048 Mbit/s (see page 65). An earlier generation of PCM equipment had 24 channels, and, like the T1 system used in the United States, operated at 1.544 Mbit/s. (In time-division multiplexing, each channel is allocated a regular timeslot, so that data from one channel is interleaved with data from all other channels on the system.) Present penetration of PCM plant is small, but it is likely to increase significantly over the next 10 years.

Trunk Circuits

Multiplexing has been a feature of the trunk network for many years, and trunk circuits are nearly always provided by means of frequency-division multiplexing (FDM) systems. (In FDM, each channel occupies a different band of frequencies within the total band of frequencies transmitted on the bearer.) Over the years bearers have changed – overhead wires have given way to coaxial cables and microwave links – but the techniques are the same. Modern coaxial systems carry over 10,000 two-way circuits per cable pair.

Table 2/1 shows the hierarchy of stages in an FDM system.

Name	Composition	Bandwidth	No of speech channels
Group	12 speech channels	48 kHz	12
Supergroup	5 groups	240 kHz	60
Hypergroup	a collection of supergroups (often 15 or 16)	variable (often 4 MHz)	variable

Table 2/1 FDM Hierarchy

All trunk circuits are 4-wire, providing separate go and return channels. Main network switching centres employ only 2-wire switching and so a transition from the 2-wire local circuit to the 4-wire trunk circuit has to be made after the MNSC. Transit exchanges employ 4-wire switching.

Digital circuits are slowly being introduced into the trunk network as part of the overall PO strategy to convert the main telecommunications network – switching and transmission – to digital working. Digital switching will come in with System X; digital transmission will be based on the 30-channel PCM system already in use. A digital network is attractive from an economic point of view, and also because ultimately it will permit speech, data and other services to be provided on a single integrated network. Table 2/2 shows the hierarchy of multiplexing stages in a time-division multiplexing environment.

Level	Aggregate bit rate	Capacity
Primary multiplex (first order)	2.048 Mbit/s	30 speech channels
Second Order	8.448 Mbit/s	4 x 30 channels
Third Order	34.368 Mbit/s	16 x 30 channels
Fourth Order	139.264 Mbit/s	64 x 30 channels

Table 2/2 TDM Hierarchy

Limitations of the Telephone Network

The requirements of a data transmission system are not the same as those for a speech transmission system, and consequently using the telephone network to carry data traffic is not without problems. The major factors encountered on the telephone network which have an undesirable effect on data are discussed below.

Attenuation (Loss)

If the attenuation (loss of signal power) on a cutomer-to-customer circuit is measured at different frequencies in the speech band, it will typically follow the U-shaped curve in Figure 2.4. Attenuation at the high end of the spectrum is caused by the frequency-dependent losses of the local ends; attenuation at the low end of the spectrum is due to the transmission bridges and other such equipment encountered on the circuit. If the circuit has been routed over an FDM system in the trunk network, some of the losses would also be attributable to the filters used to separate the channels in a group, which by design have steep cut-offs at the upper and lower ends of the speech band.

The effect of this varying attenuation is to distort the received data signal. It can be compensated for by varying the response of the receiver in a complementary fashion by means of an *equaliser*, so that the overall response of line and equaliser is approximately linear (Figure 2.5). Alternatively, if the characteristics of the circuit are already known, the transmitted data can be predistorted so as to present the correct waveform at the receiver (forward equalisation).

Group Delay Distortion

Group or envelope delay refers to the fact that not all frequencies travel at the same speed over a line. The human ear is not sensitive to this form of distortion, but its effect on data signals is severe. We saw in Chapter 1 that a square wave contains very high frequency components; delaying these frequency components by different amounts causes the square edges of a pulse to be smeared out, as Figure 2.6 illustrates. What started out as a step transition becomes spread over 500-1000 microseconds. The effect is most serious at high data rates when the fluctuations become longer than the pulse itself.

Group delay is not the same as propagation delay, which is the actual time it takes for a signal to travel from the transmitter to the receiver. Group delay only refers to the relative delay between the different frequencies that go to make up that signal. Figure 2.7 shows a typical group delay curve.

Group delay distortion is caused mainly by loaded cables and the filters used in FDM systems. It can be overcome in the same way as amplitude distortion was, by means of an equaliser which complements the characteristics of the transmission line.

Noise

White noise, which is heard as a background hiss on telephone connections, is intrinsic in any transmission system, and can never be totally removed. Fortunately it is usually of sufficiently low amplitude not to affect data signals.

Impulse noise, the presence of sudden voltage 'spikes' on a line, is more troublesome, and is the cause of most bit errors on a communications link. The electromechanical switching equipment still used in most telephone exchanges is the cause of much of this type of noise. Impulse noise, and hence bit error rate, are usually at their worst during the periods of high telephony activity (mid-morning and mid-afternoon).

PCM systems introduce a third type of noise, quantisation noise, which is a consequence of the analogue/digital conversion process (see page 65). It is not

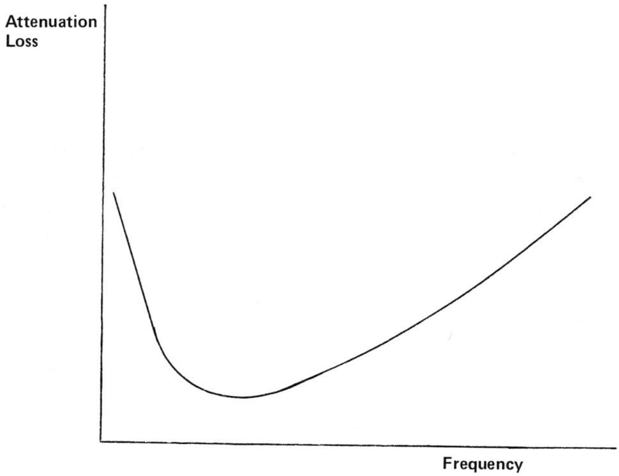

Figure 2.4 Attenuation Loss vs Frequency for a Voice Circuit

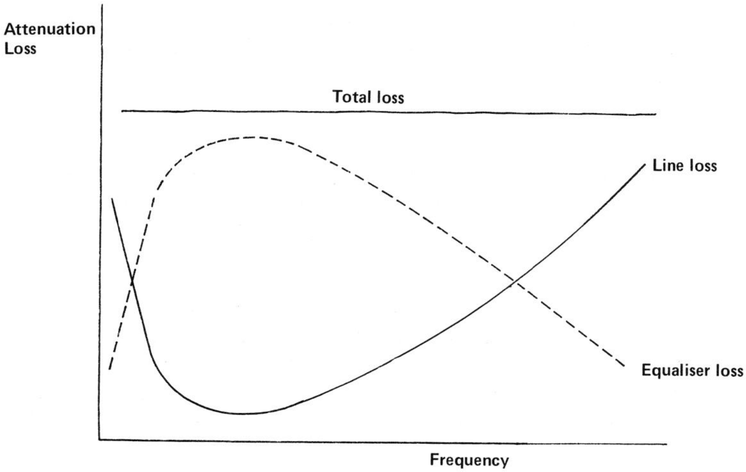

Figure 2.5 Compensating for Attenuation Loss by Equalisation

Figure 2.6 Effect of Group Delay Distortion on a Square Pulse

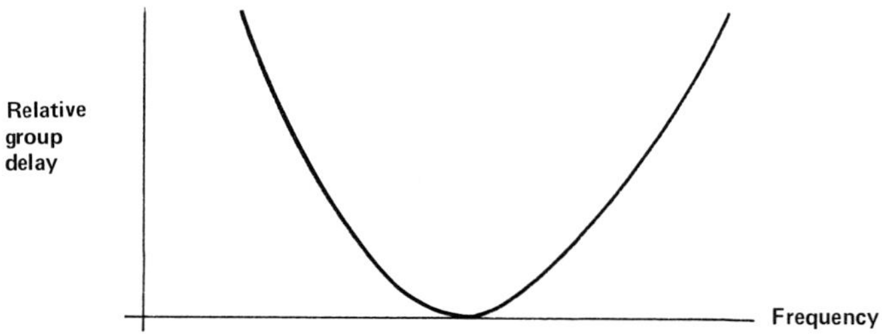

Figure 2.7 Group Delay vs Frequency Characteristic for a Telephony Circuit

significant for most users, owing to the low penetration of PCM systems at present.

Frequency Offset

The frequency division multiplexing systems used on the trunk network employ what is known as a single-sideband suppressed carrier modulation. This is described more fully in Chapter 3, but for the present it is sufficient to observe that demodulation of such a signal requires a locally-generated carrier wave which can differ in frequency from the transmitter's carrier.

If there is a difference in frequencies, the information carried over that channel will also be shifted in frequency by a corresponding amount. Fortunately there are data transmission techniques which can tolerate the shift in frequency, which normally is only a few Hertz.

Crosstalk

Crosstalk is undesirable coupling between circuits, such that the information on one circuit is 'heard' on another circuit. Crosstalk occurs between circuits which are physically close to one another. In normal circumstances it is not noticeable, but a fault condition which, for example, led to excessive power being transmitted over a circuit, could cause appreciable crosstalk.

Echo

If the 2-wire to 4-wire (hybird) transformers used on a circuit are not correctly balanced they can cause some of the signal sent on the go channel to be reflected back down the return channel (Figure 2.8). This gives rise to 'talker echo', in which the person talking hears an echo of his own voice. If the reflected signal is reflected a second time, at the sending end, the receiver gets not only the original signal but also a delayed copy of it. This is called 'listener echo'.

For data transmission, the effect is not so serious as the countermeasures. These take the form of echo suppressors which attenuate the return channel when a signal is detected in the go channel, and vice-versa. While this does prevent echoes on speech circuits, it also rules out simultaneous 2-way (full-duplex) data transmission.

Modern echo suppressors are therefore fitted with a disabling mechanism which can be activated by a modem. The suppressor remains disabled as long as data is being transmitted over the circuit. Echo suppressors are not used within the UK, but are found on long-distance intercontinental circuits.

Miscellaneous Nuisances

Sudden jumps in amplitude or phase, drop-outs (momentary losses of signal)

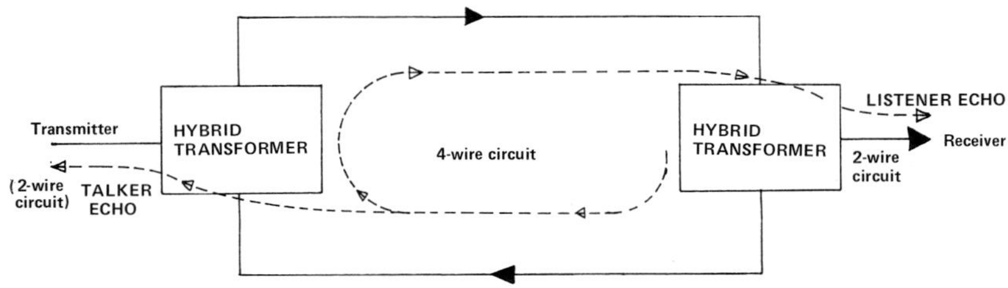

Figure 2.8 Echo Paths

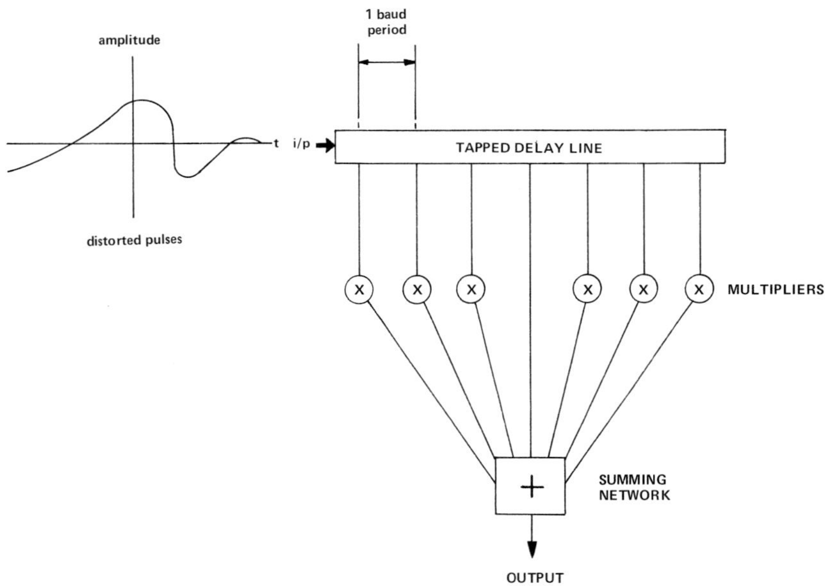

Figure 2.9 Transversal Delay Line Filter

and other such short-term impairments occur from time to time on a telecommunications circuit. They are caused by the weather, by maintenance and construction staff working on the network, and by equipment going faulty and in some cases being switched to automatic standby.

Like noise, frequency offset and crosstalk, these nuisances have to be accepted as a characteristic of the telecommunication network. Data transmission systems just have to be designed to be tolerant of them.

Steps can be taken, however, to counteract the other main impairments – attenuation and group delay distortion. As mentioned earlier, the technique is called equalisation.

Equalisation

Equalisation compensates for attenuation distortion and delay distortion. An equaliser distorts the signal in a way designed to complement the distortion introduced by the transmission line, so that the combined effect of line distortion and equalisation is to make the response of the overall system reasonably constant with frequency. Equalisation normally takes place at the receiver.

Older manual equalisers used frequency selective circuits to split up the spectrum of the incoming signal into discrete sections, each one of which could be amplified and/or delayed as appropriate.

A more common technique used today in automatic equalisers is based on the transversal delay line filter. Recall that analysis of a signal in terms of its component frequencies is only one possible approach; an amplitude-time representation is equally valid, and is the principle on which this type of equaliser is based. The equaliser examines the shape of a received pulse by sampling it at various intervals, compares that shape with that of an undistorted pulse, and works out the corrections that need to be applied to the distorted pulse to restore it to its original shape. The technique succeeds for data signals only because the shape of an undistorted pulse is already known by the equaliser.

The process can be realised in a transversal delay line filter (or its equivalent in digital circuitry), (Figure 2.9). The delay line stores incoming pulses and allows a sample to be made at each of the taps. The signal sampled at each tap can be altered relative to the centre sample by means of multipliers on the taps. The larger the number of taps the more effective is the equalisation. In operation a known data pattern is input and the multiplier coefficients are adjusted to give the best output signal. This initialisation process is known as 'training'. Equalisers which do this automatically may train initially at a lower-than-maximum data rate to reach an approximate setting, and then 'gear-shift' to a higher data rate to converge to the exact setting. Once an automatic equaliser has been set up for a circuit, subsequent training periods, following turn-round of the line for example, can be quite short. 'Adaptive' equalisers continue to adjust to changing line conditions all the time that data is being received and are usually able to tolerate short drop-outs on the line. With a non-adaptive equaliser, it is necessary to interrupt data transmission and 'retrain' if the equaliser loses synchronisation.

Equalisation is essential at high data rates when there is the need to exploit the whole of the bandwidth available on the telephone circuit. Automatic adaptive equalisation is used at data rates of 4800 bit/s and above. At 2400 bit/s, compromise equalisers are used when operating over PSTN connections. A compromise equaliser, also known as a statistical equaliser, is preset to compensate for the characteristics of an *average* PSTN connection. Below 2400 bit/s, equalisation is not required, because the transmitted signal occupies only part of the bandwidth of the telephone circuit, and the variation in attenuation and group delay over this limited band is small enough to be tolerated.

Eye Pattern

An eye pattern provides a means of assessing the quality of a data signal. The eye pattern for a 3-level signal is shown in Figure 2.10. It is a display of all the possible transitions of the data waveform superimposed upon each other. Figure 2.10 is an ideal eye pattern; in practice variations in amplitude and timing result in a less well-defined eye. The 'openness' of the eye indicates the quality of the circuit.

Similar eye patterns can be observed for 4- and 8-level signals, but it is progressively more difficult to assess the quality as the number of levels increases.

It is more common today to assess circuit quality by observing the dot pattern produced by the modulated signal. Figure 2.11 shows an ideal dot pattern for an 8-level phase-modulated signal; the quality of the circuit can be gauged from the position and sharpness of the dots. A dot pattern is in fact a signal space diagram, and will be described more fully in Chapters 3 and 4. In the USA, dot patterns may (erroneously) be called eye patterns.

PO MONOPOLY

The Post Office has a monopoly on the provision of any telecommunications facilities which are not wholly on the customer's premises. Even systems such as door intercoms in blocks of flats fall under this monopoly, but in cases like this the PO issues a general licence which permits non-PO provision.

Customers with applications not suited to the public switched networks (PSTN, PSS, Telex) can lease 'private' transmission facilities from the PO. In certain cases, the PO issues licences to specific customers, such as the gas and

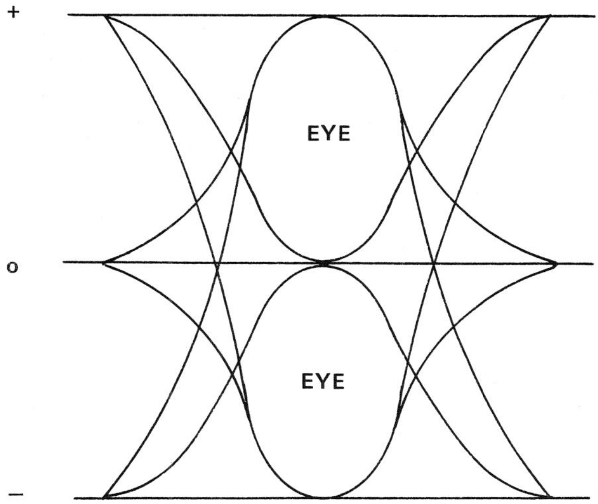

Figure 2.10 Ideal 3-Level Eye Pattern

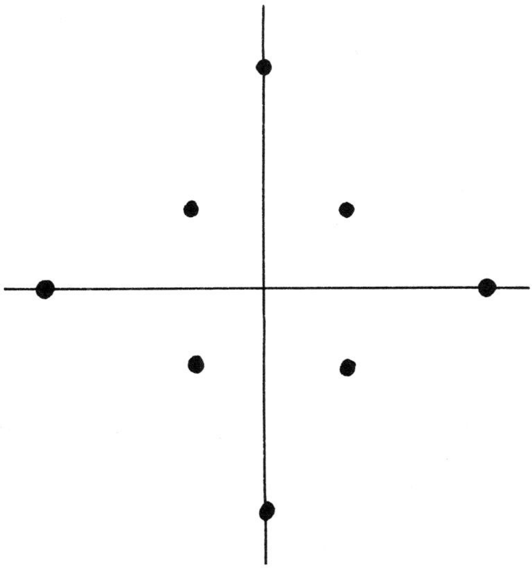

Figure 2.11 'Dot' Pattern. 8-Phase Signal

electricity undertakings, to allow them to install their own transmission plant.

PRIVATE LEASED CIRCUITS

Telegraph-type Private Circuits

Telegraph-type circuits provide a digital interface and therefore do not use a modem. There are two varieties provided by the PO – tariff H for operation up to 50 baud, and tariff J for operation up to 110 baud. Neither is widely used for data transmission. Telegraph circuits are too slow for most modern DP applications.

Telephone-type Circuits

Private leased telephone-type circuits are the most common form of private circuit. They share the same cable, amplifiers and multiplexing equipment as PSTN circuits, the only real difference being that leased circuits are permanently wired instead of being switched through exchanges. This makes them inherently more reliable, less noisy and less 'lossy' than switched connections, and of course they do not suffer from call set-up delays.

Leased circuits and PSTN connections have the same nominal bandwidth (300-3400 Hz), but the intrusion of voice frequency signalling tones used in the telephone network for supervisory purposes may limit the useful bandwidth. Figure 2.12 shows where these 'forbidden bands' are located in the spectrum. Point-to-point leased circuits on which PO signalling is not required can however ignore these limitations.

The PO offers both 'scheduled' and 'unscheduled' circuits. (In the USA, the corresponding terms are 'conditioned' and unconditioned' lines.) Unscheduled circuits are specified only in terms of their nominal loss (in decibels) at 800 Hz. Table 2/3 lists them.

Loss at 800 Hz	Presentation
17dB	2-wire
10dB	2-wire
3dB	2-wire
0dB	4-wire

Table 2/3 PO Unscheduled Circuits

Scheduled circuits are much more closely specified; Table 2/4 lists the 4 categories. Schedule D circuits are similar to the old PO tariff T circuits, and conform to CCITT recommendation M.1020 for high quality international

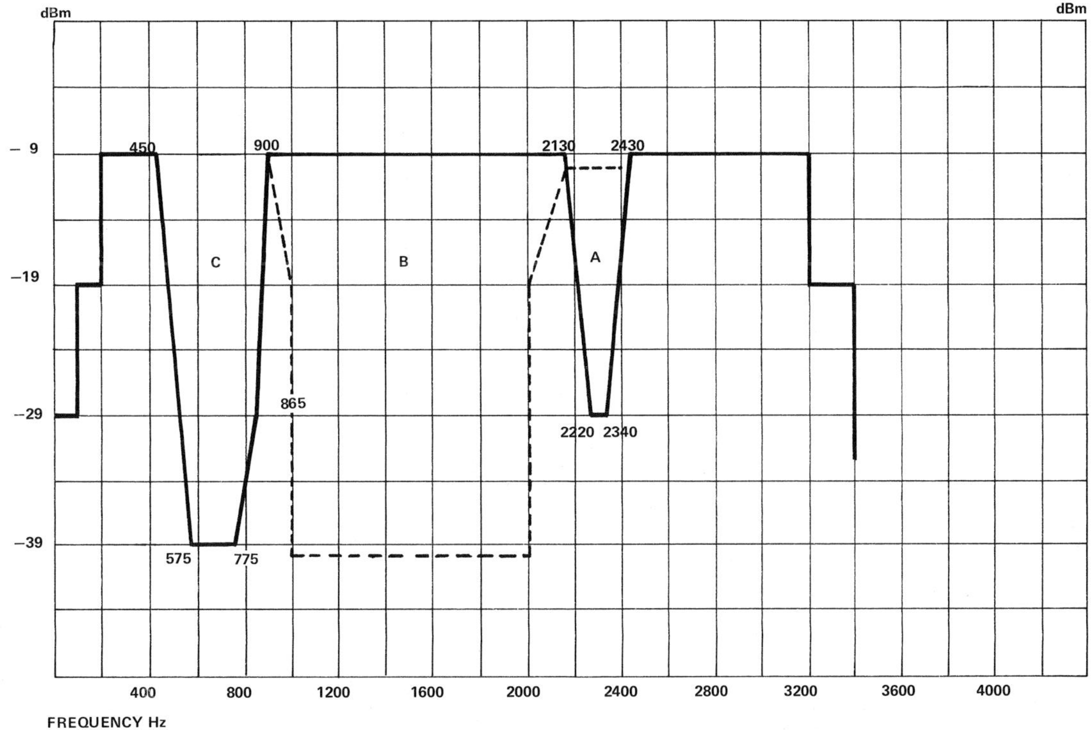

NOTES: (1) Signals are permitted in Area A only if accompanied by signals in Area B at a power level not lower than 12 dB below the power level of the signal in Area A.

(2) Signals are permitted in Area C provided that there is no false operation of trunk signalling equipment (SSAC 1).

Figure 2.12 Permitted Power Levels on UK Telephony Circuits

leased circuits.

Short-distance Circuits

Circuits less than 7 miles long will often only involve one local exchange, and so the circuit takes the form of a physical pair of wires throughout. DC and baseband transmission will normally be possible. DC transmission may also be possible on circuits up to 25 miles, since up to this distance physical pairs are normally available in the form of junction circuits, but dc paths cannot be provided for circuits longer than 25 miles. Baseband transmission over about 7 miles is only possible if unloaded junction circuits exist. In fact any transmission over speech band circuits which is outside the range 300-3000 Hz is treated by the PO as non-standard, and is subject to special negotiation.

Wideband Circuits

Wideband circuits are available at 48 kHz, 240 kHz, and 5.5 MHz. All these services require special cable to be laid between the customer's premises and the nearest main network access point, such as a repeater station, and can thus take some time to obtain. The 48 kHz circuits are used for 48 kbit/s data links (or up to 168 kbit/s using proprietary modems); the other circuits are employed in the main for multiplexed speech and video rather than data.

Circuit Configurations

The PO offers three basic configurations: point-to-point circuits, omnibus circuits, and multipoint circuits (Figure 2.13). Point-to-point circuits connect together two terminals. Omnibus circuits connect three or more terminals with complete intercommunication between each. Multipoint circuits connect one central station to up to 12 outstations; intercommunication between outstations is not possible.

The multipoint configuration is designed for data applications only and is well suited to the hierarchical structure of most computer networks. The other two configurations can be used for data or speech.

Presentation of Private Circuits

Circuits are presented to the customer as either 2-wire or 4-wire circuits. A 2-wire circuit uses the one pair for transmission and reception; a 4-wire circuit uses one pair for transmission and the other for reception. Schedule A circuits are 2-wire; the others may be 2-wire or 4-wire. Omnibus and multipoint configurations are 4-wire. Telephone-type circuits have a nominal impedance (ie resistance at audio frequencies) of 600 ohms.

Table 2/4 Parameters of PO Scheduled Circuits

	Schedule **A**	Schedule **B**	Schedule **C**	Schedule **D**
Loss/Frequency Response (dB relative to 800 Hz)				
300-500 Hz	−7 to +12	−3 to +10	−2 to +7	−2 to +6
500-2000 Hz	−7 to +8	−3 to +6	−1 to +4	−1 to +3
2000-2600 Hz	Not specified	−3 to +6	−1 to +4	−1 to +3
2600-2800 Hz	Not specified	−3 to +10	−1 to +4	−1 to +3
2800-3000 Hz	Not specified	−3 to +10	−2 to +7	−2 to +6
Group Delay/Frequency Response μ secs relative to minimum				
500-600 Hz	Not specified	Not specified	Not specified	3000
600-1000 Hz	Not specified	Not specified	Not specified	1500
1000-2600 Hz	1250	1000	1000	500
2600-2800 Hz	Not specified	Not specified	Not specified	3000
Random Circuit Noise (dBmOp)	−42	−42	−42	−45
Impulsive Noise No more than 18 Impulse Noise counts to exceed the threshold limit in any period of 15 mins	Threshold Limit −18 dBmO	Threshold Limit −18 dBmO	Threshold Limit −18 dBmO	Threshold Limit −21 dBmO
Signal to Quantising Noise Level (dB)	22	22	22	22
Maximum Frequency Error (Hz)	2	2	2	2
Transmit to receive Crosstalk Attenuation 4 wire Presented Circuits measured at 2000 Hz (dB)	Not applicable	45	45	45
Signal to Listener Echo Ratio (dB) 2 wire Presented Ccts	16	20	20	20
Variation with time of the Insertion Loss at 800 Hz (dB)	±3	±3	±3	±3

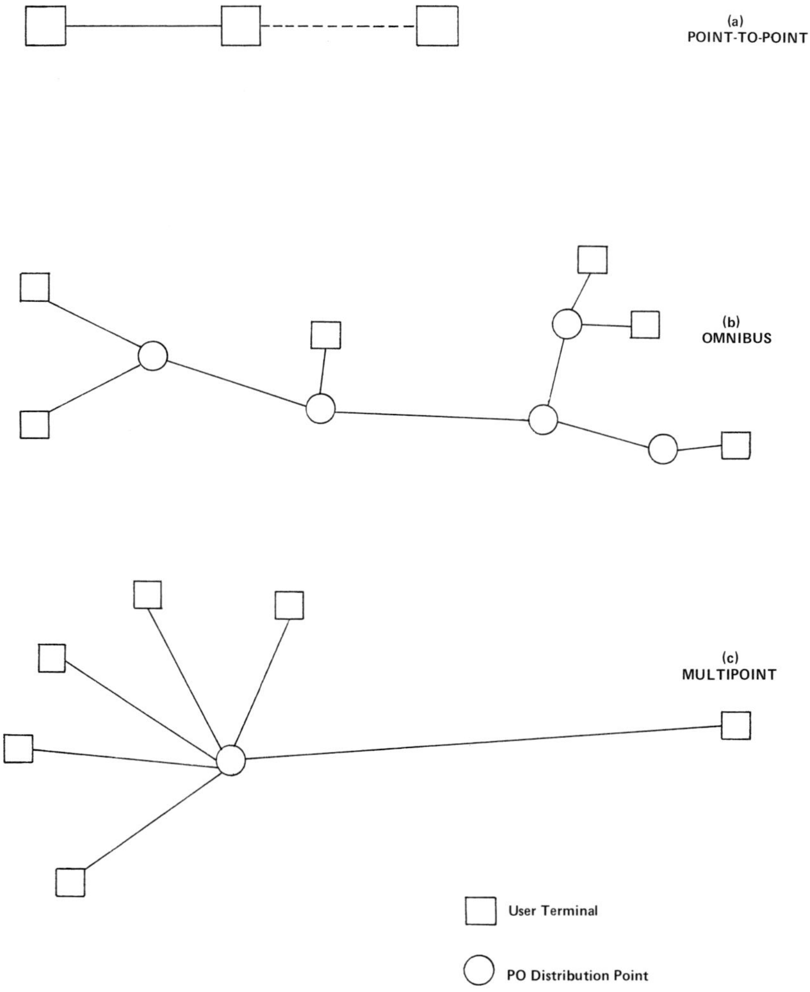

Figure 2.13 Circuit Configurations

(a) Point-to-point (connections may be in tandem)

(b) Omnibus

(c) Multipoint

ATTACHMENTS TO PO EQUIPMENT

The PO maintains strict control over the equipment connected to its own plant and equipment. This control is designed to safeguard PO personnel and the operation of the PO network.

Normally modems used on the PSTN have to be supplied by the PO, although this rule is relaxed occasionally when the PO is unable to provide a modem of the right type. The rule has also been relaxed in the case of viewdata television sets, which have integral modems. Any terminal connected to a PO modem has to be 'type approved'. As well as safety checks, the approval procedure for data terminals includes a check that the terminal conforms to the relevant interface standards, both electrically and procedurally. The supplier also tells the PO which Datel service codes (indicating the options to be strapped up in the modem) are to be released for this particular terminal. A list of Permissible Attachments is held in every PO telephone area sales (special services) department.

Approval of equipment connected to private leased circuits is less exhaustive, and involves checking that the equipment is safe and that it will not deliver excessive power to line. In the case of modems for use on private circuits, the PO only examines the analogue line interface; it is the customer's responsibility to check the digital terminal interface.

3 Modems and Modulation Theory

MODEMS

A modem performs two main functions, modulation and demodulation (wherein lies the origin of the word modem).

Inside, a modem comprises three sections: circuitry associated with the transmit (modulation) function, circuitry associated with the receive (demodulation) function, and circuitry for timing and power supplies which serves the other two.

Figure 3.1 is a simplified block diagram of a modem. We shall not delve any deeper than this into the inner workings of a modem, because this book is essentially for people who use modems, and the important thing to the user is the behaviour of a modem as a unit within a total data communications system.

The following chapters are therefore devoted to how the modem looks to the components on either side of it – the terminal or computer on one side, and the transmission line on the other. We start with some basic theory.

MODULATION THEORY

Put very simply, modulation is a process whereby a high-frequency wave is made to carry a lower-frequency wave. There are many instances where a transmission medium will convey high-frequency signals but not low-frequency signals, and this is when modulation is required. Radio provides a good example. High-frequency electromagnetic waves propagate well through space, but low-frequency speech and music signals do not. In radio, therefore, the speech or music signals are allowed to modulate a high-frequency carrier of several hundred kilohertz, and the modulated high-frequency signal is then broadcast. At the radio receiver this modulated high-frequency wave is demodulated to retrieve the original speech or music signal.

'High' and 'low' frequencies are relative terms. We have already seen that telephone circuits transmit 'high' frequencies between 300 and 3400 Hz, but

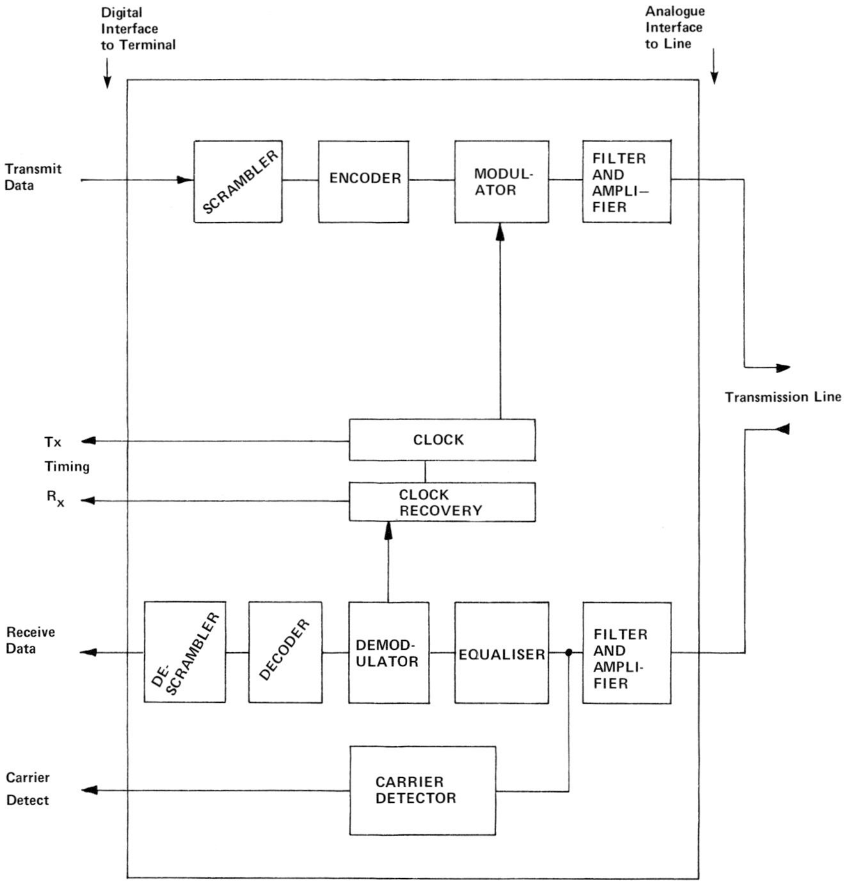

Figure 3.1 Block Diagram of a Typical Modem

not 'low' frequencies below 300 Hz. Data signals contain low-frequency components less than 300 Hz, and thus modulation is required if these signals are to be transmitted over telephone circuits.

The three attributes of a carrier that can be altered are amplitude, frequency and phase. Figure 3.2 shows a sine wave. The meaning of amplitude is obvious – it is the size (voltage) of the wave. Frequency is the number of cycles per second and is measured in Hertz. Phase is a more difficult concept to appreciate, and it is worth spending some time on explaining just what phase is.

Our starting point is a phasor or vector diagram; this is just a pointer, initially pointing due East, which rotates anticlockwise. It is analogous to the hand of a clock (Figure 3.3a). Imagine the pointer rotating slowly anticlockwise; if we were to plot a graph of the angle of rotation against the distance of the end of the pointer above (+) or below (–) the West-East baseline, we would obtain a sine wave as the sketches in Figure 3.3b reveal. As the pointer continued to rotate, the wave would be replicated, each revolution corresponding to one cycle. The number of revolutions per second corresponds to the frequency of the wave, the length of the pointer to its amplitude (constant in this case), and the angle of rotation to the phase of the wave. One complete cycle of the wave corresponds to a phase change of 360° (2π radians). The phase of the wave is continually increasing as the wave propagates; after say 1¼ cycles, the true phase is 450°. In a practical system the true phase is rarely known; the important factor is the phase within one cycle, ie 90° in the case of 1¼ cycles.

The remainder of the chapter covers the three different types of modulation.

Amplitude Modulation

In an amplitude modulation (AM) system, the modulating waveform is allowed to modify the amplitude of a high-frequency carrier wave. Figure 3.4 shows the waveform at the input and output of a modulator for the simple case where the modulating waveform is a single frequency tone (f_m). An analysis of the frequencies present in the output signal reveals the presence of not only the input frequencies f_c and f_m, but also the sum and difference frequencies $f_c + f_m$ and $f_c - f_m$. This is not immediately apparent from the output waveform in Figure 3.4, and a better representation is to use a horizontal axis showing frequency instead of time, as in Figure 3.5.

If the modulating waveform is a complex waveform such as a speech signal which contains many frequencies, the various sum and difference frequencies occupy two bands, one above and one below the carrier frequency (Figure 3.6). They are known as upper and lower sidebands. The upper sideband is a replica of the original speech signal, just transposed up in frequency

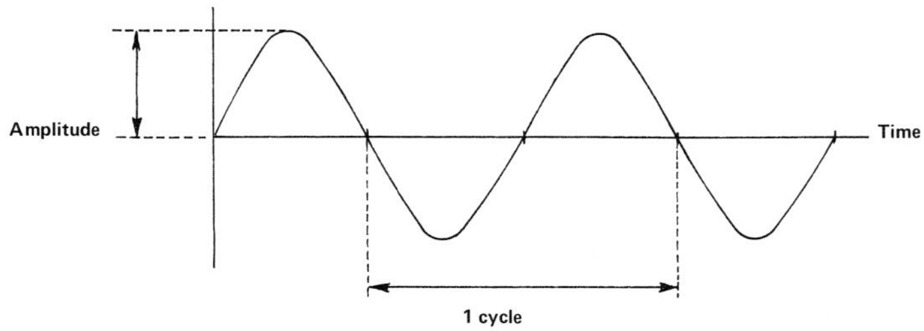

Figure 3.2 Sine Wave

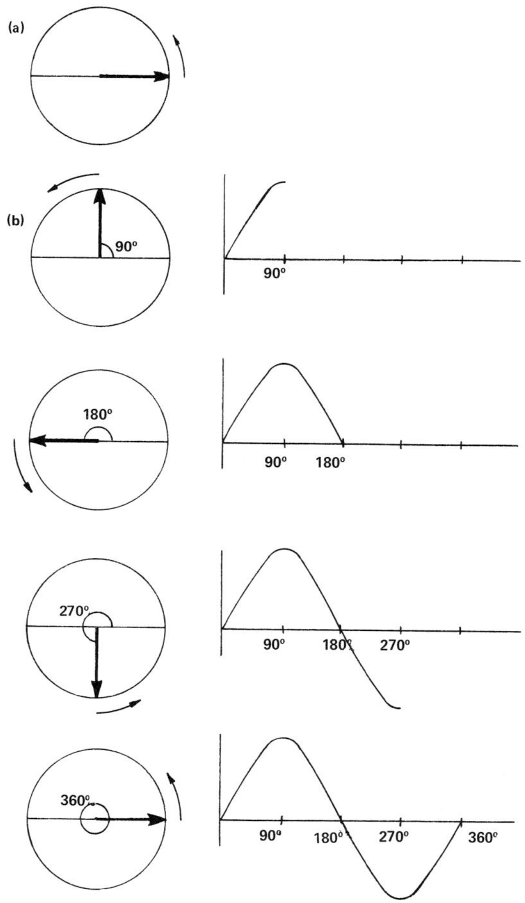

Figure 3.3 (a) Phasor Diagram (b) Generation of a Sine Wave

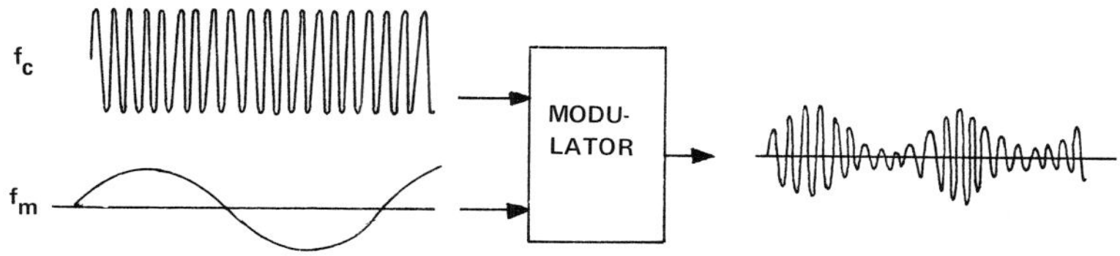

Figure 3.4 Amplitude Modulation

f_c = carrier f_m = modulating waveform

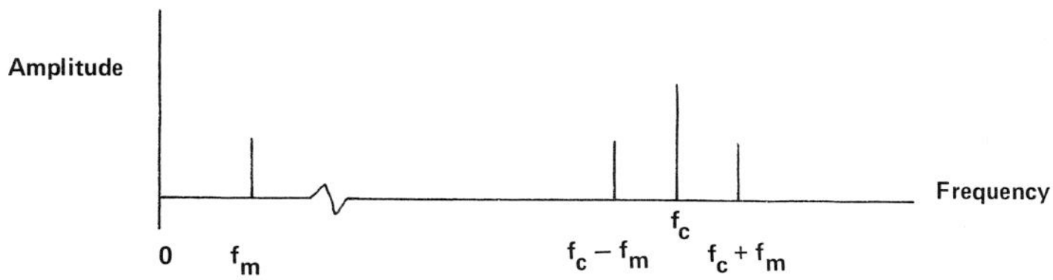

Figure 3.5 Frequency Components of an AM Signal as in Figure 3.4

by f_c Hz. The lower sideband is an inverted replica of the original signal ie the highest frequencies in the original are the lowest frequencies in the lower sideband. The lower sideband is a mirror image of the upper sideband as reflected in the carrier frequency f_c. (The triangular shape of the speech signal in Figure 3.6 is merely a convention which illustrates the inversion of the lower sideband; it does not represent the spread of energy at different frequencies within the signal.)

An AM system which transmits both sidebands and the carrier is known as a double sideband (DSB) system. The carrier contains no useful information and is sometimes suppressed, but with or without carrier the bandwidth of a DSB system is still twice that of the original signal. To reduce the bandwidth it is possible to suppress not only the carrier but also one of the two sidebands, since they both carry the same information. This is known as single-sideband suppressed carrier (SSB-SC) operation; the technique is used extensively in FDM telephony multiplexing systems. Effectively all SSB-SC does is to provide a new signal identical to the original, but transposed upwards in frequency. The carrier frequency is chosen to give the desired transposition.

Demodulation of an AM signal is achieved by mixing the modulated signal with a carrier of the same frequency as that used for modulation. The original signal then emerges as a difference frequency (or band of frequencies) and can be filtered out from the other signals present. In SSB-SC operation, the carrier used for demodulation is generated locally, and may not be 'locked' in any way to the frequency of the carrier used for modulation. Small differences between the two frequencies are the cause of the frequency offset encountered on telephone circuits.

Amplitude Modulation Using Digital Signals

Figure 3.7b illustrates a carrier wave modulated in amplitude by the binary data signal shown in Figure 3.7a. A special case of amplitude modulation is when the lower of the two amplitude levels is reduced to zero; the modulation process then reduces to switching the carrier on and off (Figure 3.7c). However, the variation in transmitted energy makes this technique unsuitable for data transmission over telecommunications networks.

A square wave like the one in Figure 3.7a contains high-frequency components, and in a practical AM system the data signal would be passed through a low pass filter prior to modulation. This rounds off the square wave (Figure 3.8), but does not affect the information content of the data signal. The output of the modulator contains frequencies as shown in Figure 3.9. Note that because the binary data signal extends down to zero frequency, the upper and lower sidebands actually meet at f_c. This makes it difficult to suppress the carrier, or to suppress one sideband and the carrier, without affecting the other sideband. What can be done to reduce the bandwidth of the modulated

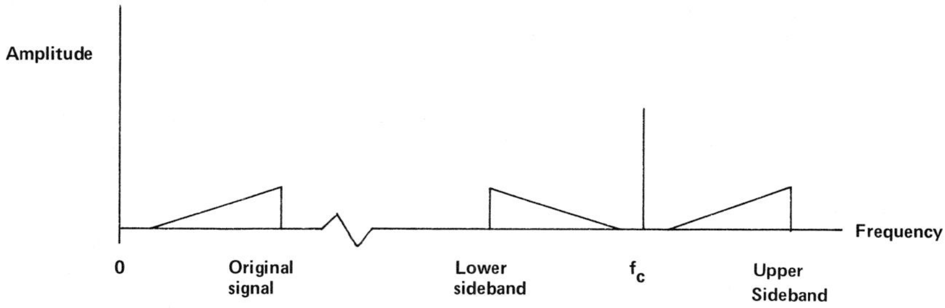

Figure 3.6 Upper and Lower Sidebands in AM

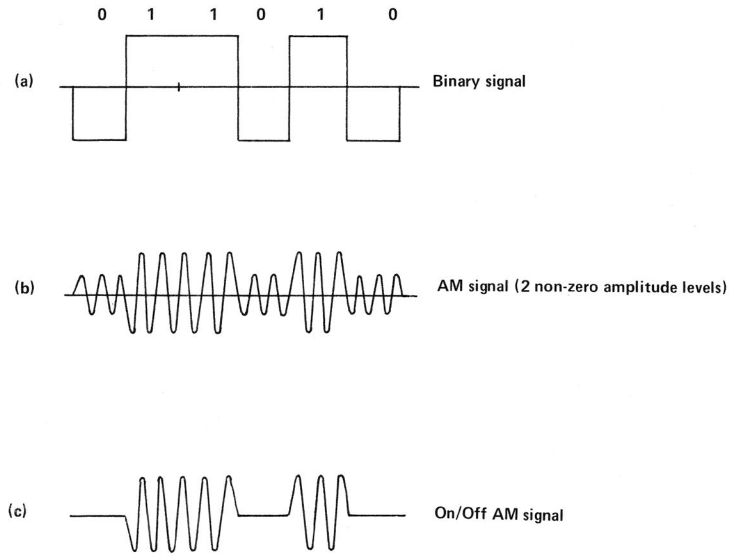

Figure 3.7 Amplitude Modulation with Binary Digital Modulating Signal

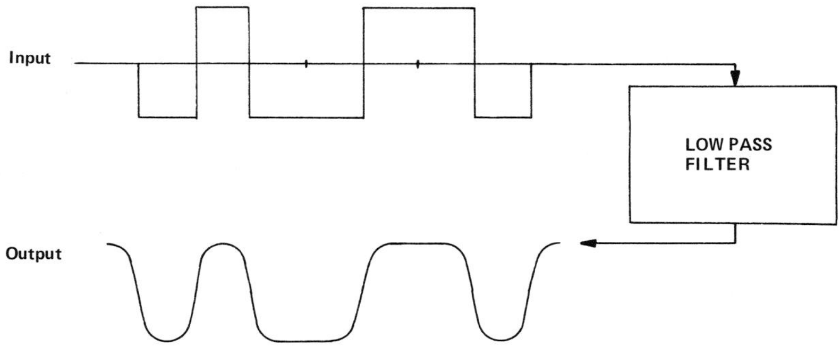

Figure 3.8 Rounded Square Wave Produced by a Low Pass Filter

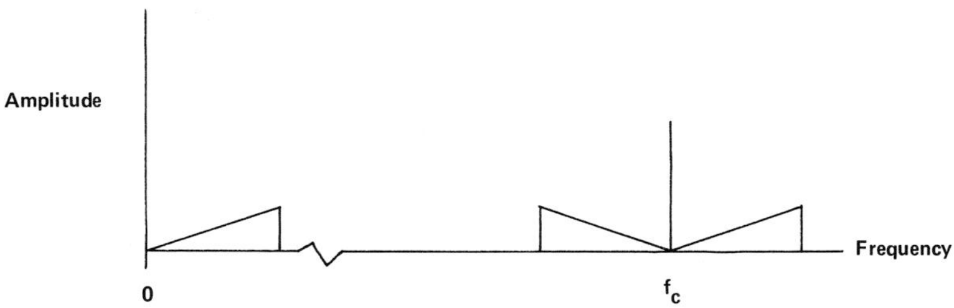

Figure 3.9 Sidebands in an AM Signal with Baseband Modulating Waveform

signal is to suppress most of one sideband, leaving only a vestige of it near the carrier frequency (Figure 3.10). There is no loss of information since the lower sideband merely duplicates the information in the upper sideband. The technique is called vestigial sideband (VSB) modulation.

With clever filter design it is possible to suppress the carrier in VSB systems. It leads to part of the upper sideband being suppressed, but the vestige of the lower sideband which remains supplies the missing frequencies.

True single-sideband amplitude modulation with a digital modulating signal can only be achieved by scrambling the original data (ie randomising the bit stream) in order to remove low-frequency components caused by long strings of 1s or 0s. This has the effect of separating the sidebands from the carrier (Figure 3.11), thus making it feasible to filter out one sideband and the carrier.

Pulse Amplitude Modulation (PAM)

Pulse amplitude modulation, when used with a digital modulating waveform, provides a means of coding more than one bit per baud, by encoding the binary data signal as a signal with more than two levels (sometimes called an m-ary signal).

For example, the bits of a binary data signal could be sampled in pairs. There are four possible combinations of a pair of bits, and thus each pair could be encoded as one of four amplitude levels. The encoded 4-level signal has half the baud rate of the original data signal, and can be used to amplitude-modulate a carrier in the usual way.

Frequency Modulation

In frequency modulation (FM) systems, the *frequency* of the carrier is altered in sympathy with the modulating waveform. Systems in which the modulating waveform is a binary signal, so that the carrier is switched abruptly from one frequency to another, are referred to as frequency shift keying (FSK) systems. Figure 3.12 illustrates a binary data signal and the corresponding frequency-modulated carrier.

Analysis of an FM signal is not as easy as it was for an AM signal, but one of the simplifications that can be made is to regard the FM signal as the sum of two AM signals, as Figure 3.13 demonstrates. It is not then surprising to find that the bandwidth required for an FM signal is up to twice that required for an AM signal. Frequency modulation is superior to amplitude modulation in terms of its tolerance to certain of the impairments found on the telephone network, and so it tends to be used at low data rates where the limited bandwidth available is not a serious restriction. FSK is an asynchronous modulation technique; there is no need for a clock in the modem.

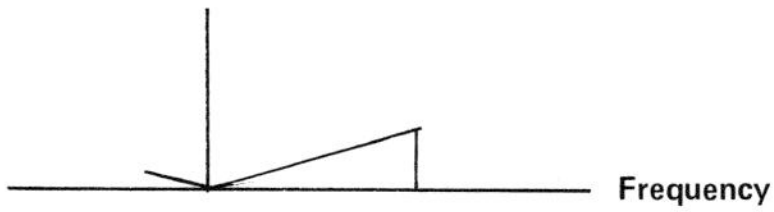

Figure 3.10 Vestigial Sideband AM

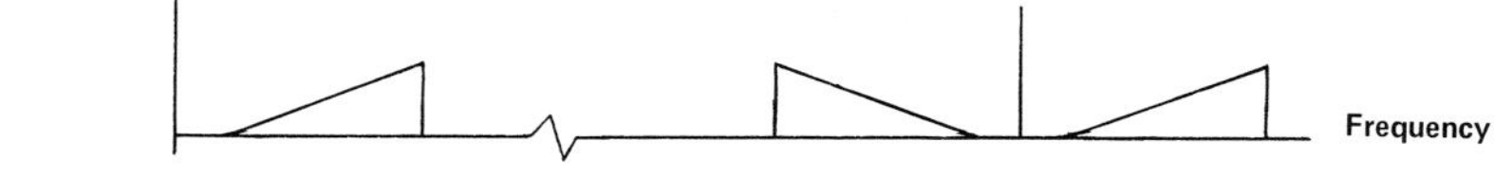

Figure 3.11 Sidebands in an AM Signal with Scrambled Baseband Modulating Signal

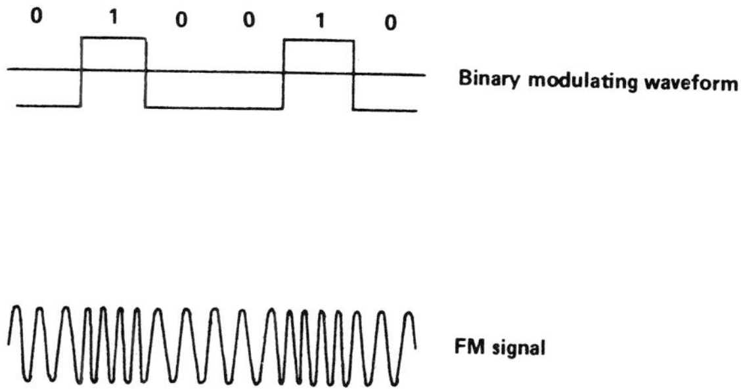

Figure 3.12 Frequency Shift Keying (FSK)

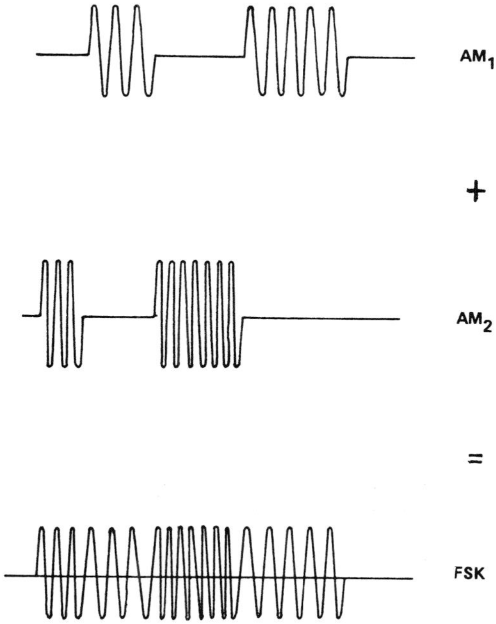

Figure 3.13 FSK as the Sum of Two AM Signals

Phase Modulation

The simplest type of phase modulation is known as phase shift keying, by analogy with the FSK system just described. Imagine two oscillators both generating a sine wave at the carrier frequency, but 180° (half a cycle) out of phase. One oscillator is connected to line whenever there is a 0 in the data signal, and the other whenever there is a 1. The waveforms are shown in Figure 3.14.

If the receiver has a reference carrier against which to compare the phase of the incoming signal, it will be able to demodulate the signal. However, the receiver must be given some indication at the start of transmission as to which phase represents a '0' and which a '1'.

A modulation system in which a carrier is generated locally at the receiver and used to demodulate the incoming signal is known as a *coherent* system. The phase modulation system just described is a coherent system. It is also known as fixed reference phase modulation.

Another technique is differential phase modulation. Each bit in the data signal is coded as a phase change relative to the previous phase of the carrier. For example, a '0' bit could be coded as a 90° phase change, and a '1' bit as a 270° phase change. Figure 3.15 shows what the modulated waveform would look like.

By comparing the phases of adjacent signal elements in the incoming signal and determining whether the phase change is 90° or 270°, the receiver can reconstruct the original data signal unambiguously. No fixed reference phase is needed in this system, but in fact some differential coding systems do use a reference signal to give improved performance.

Phase modulation using a binary modulating signal can be considered as the difference between an amplitude-modulated wave and an unmodulated carrier, as Figure 3.16 demonstrates. Put another way, a phase-modulated signal is equivalent to a suppressed carrier double-sideband amplitude-modulated signal. Phase modulation thus requires twice the bandwidth of the original data signal.

Multi-Phase Modulation

Two-phase modulation as described above codes one bit of data per phase change. With more than two phases, it would be possible to code more than one bit per phase change, and thus to increase the bit rate without altering the modulation rate. To code two bits per phase change, for example, we would need four possible phase changes to represent the four combinations of two bits (00, 01, 10, 11).

Using differential modulation, the technique would be to divide the data

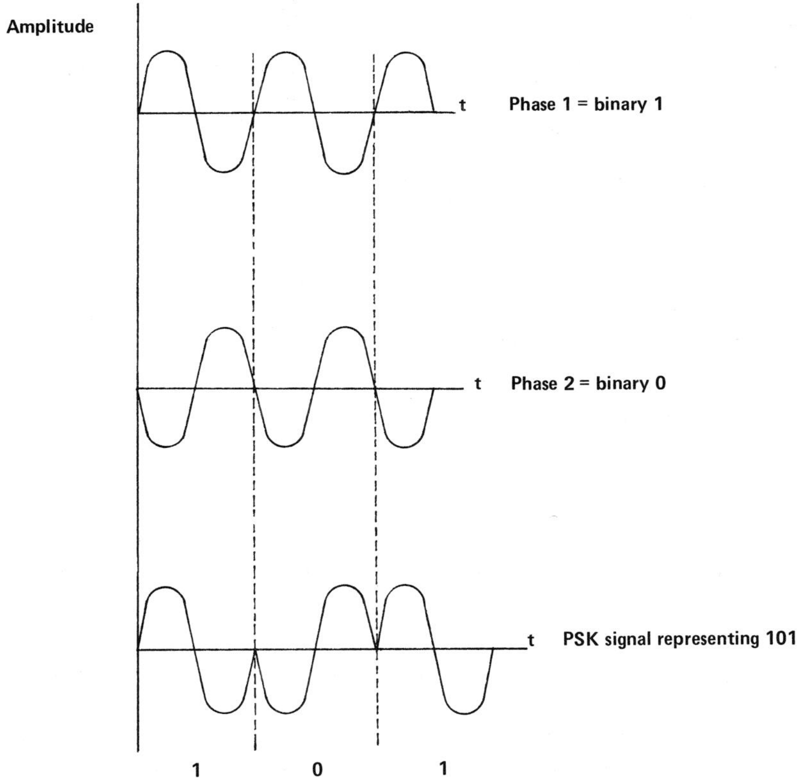

Figure 3.14 Phase Shift Keying

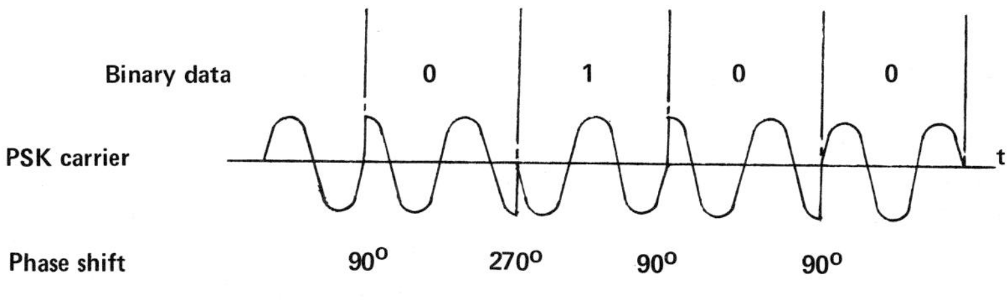

Figure 3.15 Differential 2-phase Encoding

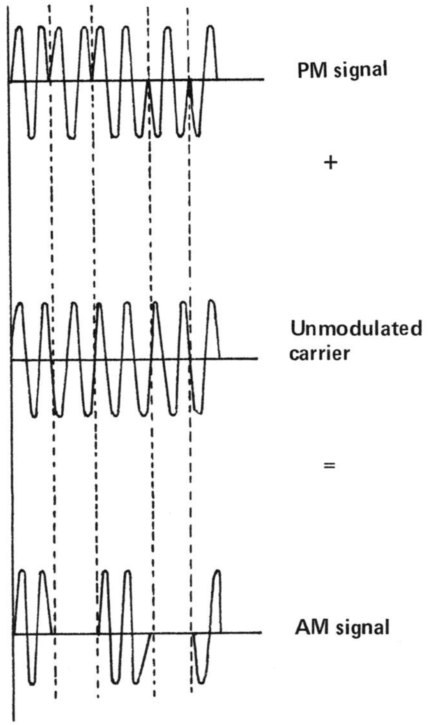

Figure 3.16 Relationship between PM and AM

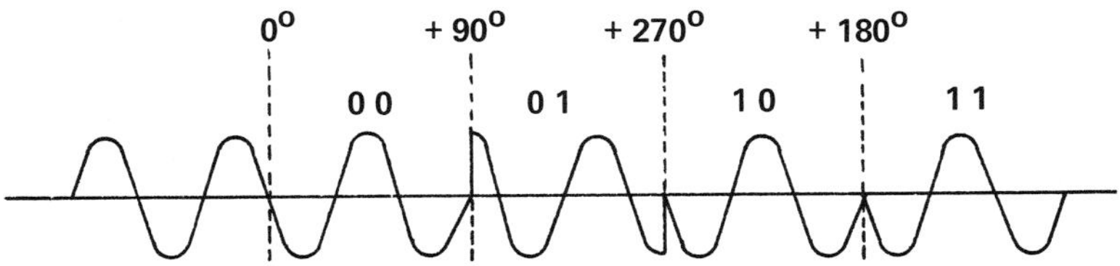

Figure 3.17 4-phase Modulated Signal

signal into pairs of bits ('dibits'), and to shift the phase of the carrier signal in one of four ways, according to the dibit combination. Table 3/1 shows one possibility.

Dibit	Phase change
00	0°
01	90°
11	180°
10	270°

Table 3/1 Coding of dibits

The phase changes shown in Table 3/1 are phase shifts relative to the previous phase of the carrier. Figure 3.17 gives an example of a carrier modulated in this way.

An alternative method of presenting the information given in Table 3/1 is to use a type of phasor diagram showing the possible phase shifts (Figure 3.18). A diagram of this sort provides a very convenient method of illustrating the states that a transmitted carrier can occupy, especially for the more complicated phase- and amplitude-modulation techniques.

We have already seen that, for a binary modulating signal, two-phase modulation is equivalent to double-sideband suppressed carrier (DSB-SC) amplitude modulation. Four-phase modulation is equivalent to two DSB-SC waves with carriers 90° out of phase with each other, being transmitted simultaneously, and can be considered as a special type of quadrature amplitude modulation (QAM). Phase modulation is a synchronous modulation method, requiring a clock in the modem.

Phase- and Amplitude-Modulation (Quadrature Amplitude Modulation)

It is possible to combine phase modulation and amplitude modulation to give a further increase in the number of bits per baud. Figure 3.18 showed the possible phase shifts in one type of 4-phase system coding two bits/baud. If now the amplitude of the carrier was allowed to adopt one of two possible levels for each of these phase shifts, there would be eight possible states which the carrier could adopt each baud period (Figure 3.19). This would allow three bits per baud to be carried.

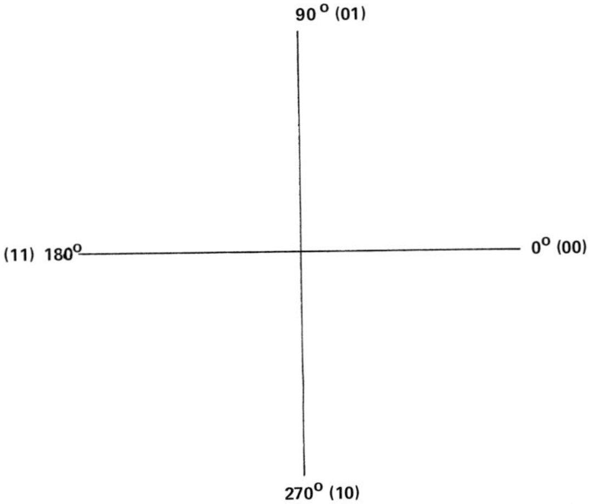

Figure 3.18 Phase Diagram showing Phases in 4-phase Modulation

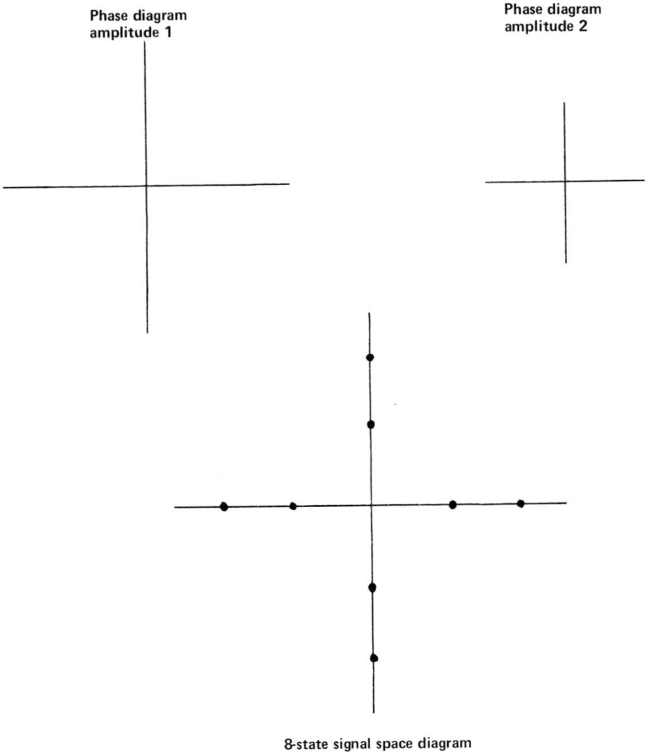

Figure 3.19 Phase-Amplitude-Modulation (4 phases, 2 amplitudes)

Multicarrier Amplitude- Phase-Modulation

One of the newest amplitude-phase-modulation techniques relies on the simultaneous transmission of multiple carriers. One particular implementation employs 48 carriers, separated by 45 Hz spacings. By a combination of phase- and amplitude-modulation, each carrier can occupy one of 32 discrete states each baud period, permitting five bits per baud to be carried. Thus the 48 carriers can carry 5 x 48 = 240 bits per baud. For operation at 9600 bit/s, the modulation rate need only be 40 bauds; such a slow rate is very tolerant of the phase and amplitude hits so common on the telephone network. The actual bandwidth used is 2240 Hz. Modulation and demodulation are all performed digitally, in a microprocessor.

The technique illustrates the extent to which cheap electronics now makes it possible to realise ideas that would never have been practicable a while ago.

Pulse Code Modulation

Pulse Code Modulation (PCM) is included in this discussion of modulation systems because of its importance. It is normally encountered as a system for transmitting analogue signals such as speech in a digital format; it is not a modulation technique used by modems, (although some baseband modems do employ the same HDB3 line code as PCM systems – see below).

In PCM, the analogue signal is sampled at a rate at least twice that of the highest-frequency component of the analogue signal. The PCM systems used on the telephone network sample 8000 times a second. Each sample yields a voltage level which is coded into a seven-bit code. Coding follows a logarithmic law, to better represent the wide variation in possible speech levels. These seven bits, together with an eighth bit to signify the sign of the signal, constitute an octet. The bit rate of a single PCM channel is thus 8 x 8000 = 64 kbit/s.

The standard telephony PCM systems use time-division multiplexing to carry 32 octet-interleaved channels (30 speech channels + two control channels) at an aggregate rate of 2.048 Mbit/s.

The 2.048 Mbit/s binary stream is not suitable for transmission directly to line, and so it is converted to a three-level (ternary) line code known as HDB3 (*High Density Bipolar 3*). HDB3 coding ensures that the transmitted data has no net dc component and that transitions are sufficiently frequent to permit reliable clock extraction. Figure 3.20 shows the stages in HDB3 coding.

At the receiver, the bit stream is decoded back into binary and demultiplexed; the samples are reconstituted from the coded bits to produce a 'quantised' waveform very similar to the original (Figure 3.21). This quantisation distortion is not noticeable in telephony; in data transmission it is manifested as another source of noise.

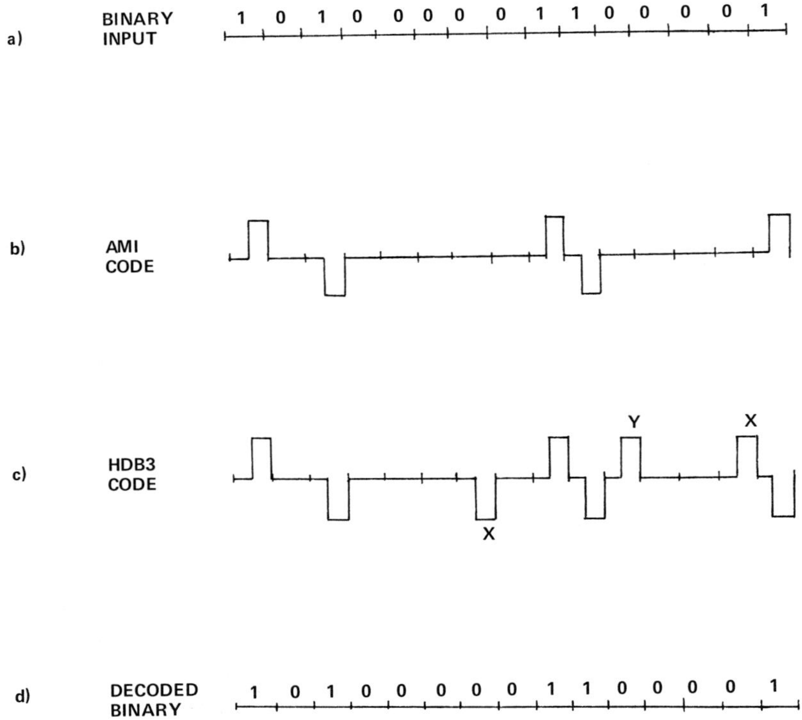

a) Binary data input

b) Alternate Mark Inversion (AMI)

– binary 1s in data converted to alternate positive or negative pulses

c) HDB3. Certain zeros replaced by ones to ensure

– ≯ 3 consecutive zeros in transmitted data (hence X pulses)

– successive X pulses are of opposite polarity (hence Y pulse)

X pulses violate the AMI principle, and can thus be identified at receiver

d) decoded binary data recovered by receiver

Figure 3.20 HDB3 Coding

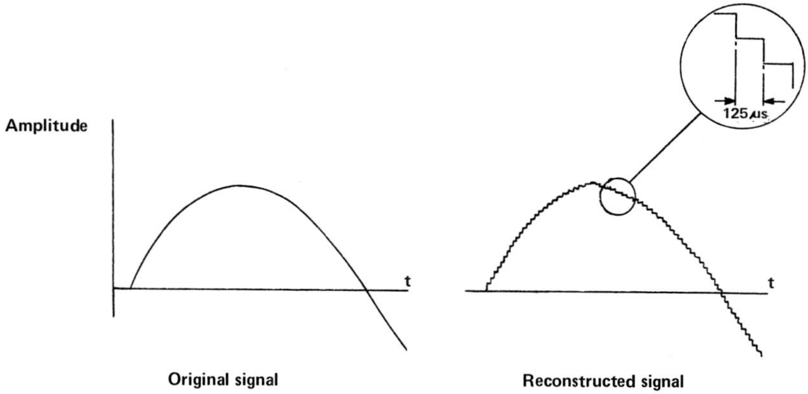

Figure 3.21 Pulse Code Modulation (PCM)

4 Standard Modems

CCITT MODEMS

CCITT stands for *Comité Consultatif International Télégraphique et Téléphonique*; it is part of the International Telecommunications Union which in turn is a specialist agency of the United Nations Organisation. CCITT is the standardisation body of the PTTs and common carriers, and also includes representatives from telecommunications equipment manufacturers. CCITT issues recommendations, not standards. European Administrations have tended to follow these recommendations fairly closely (but not exclusively) and in Europe many have the authority of international standards. The United States has not in the past been constrained by these recommendations, but today, in the field of data communications at least, US manufacturers are conforming to CCITT recommendations to a greater extent than ever before.

CCITT has drawn up recommendations for a range of modems from 200 bit/s to 48 kbit/s and these are described in the following pages. Appendix 1 gives further information on CCITT terminology. PTTs and common carriers usually provide modems conforming to CCITT recommendations, but may also provide non-standard modems for purely domestic use.

200 Baud Modem CCITT Rec V.21 (1964)

Nearest equivalents: *UK PO Modem 2,21*
US Bell 103,113

V.21 modems employ frequency shift keying (FSK), and provide full-duplex (both ways simultaneous) operation over the PSTN at up to 300 bauds. FSK is an asynchronous modulation technique; the receiving modem does not extract timing signals from the incoming signal in order to demodulate the data. The modem is in fact completely transparent, able to accept data in any code, at any data rate up to its maximum. This makes it well suited to the variety of codes and speeds used by low-speed data terminals.

The frequencies used are shown in Table 4/1.

	0, Space	1, Mark
Channel 1	1180	980 Hz
Channel 2	1850	1650 Hz

Table 4/1 V.21 Frequencies

Conventionally on PSTN calls, the modem which receives an incoming call transmits on channel-2 frequencies, although this feature can usually be overridden if required.

The Bell modems use different frequencies (channel 1: 1070 Hz = 0, 1270 Hz = 1. Channel 2: 2025 Hz = 0, 2225 Hz = 1).

600/1200 Baud Modem CCITT Rec V.23 (1964)

Nearest equivalents UK PO Modem 1,20,22
US Bell 202

Frequency shift keying is employed, but the frequencies differ from those used at 200 baud. Transmission at 1200 baud can be full-duplex over 4-wire leased circuits, but only half-duplex over PSTN connections owing to bandwidth limitations. (Half-duplex implies transmission in either direction, but not in both directions simultaneously).

The standard frequencies are shown in Table 4/2.

	0, Space	1, Mark
Transmission up to 600 baud	1700	1300 Hz
Transmission up to 1200 baud	2100	1300 Hz

Table 4/2 V.23 Frequencies

There is a certain amount of bandwidth available at the lower end of the spectrum, and this is exploited to provide an optional low-speed 75 baud secondary channel, again using FSK. The frequencies are:

0	1
450 Hz	390 Hz

This channel operates in the opposite direction to the main channel, but simultaneously with it. Operation with a secondary backward channel is termed asymmetric duplex. When not employed for user data, the backward channel is sometimes employed for diagnostic purposes.

The V.23 modulation technique is essentially asynchronous, but is is also possible to use the modem in a synchronous fashion if required by installing a clock to provide timing for the data terminal.

The Bell modem 202 is not compatible with the V.23 modem because it uses different frequencies (2200 Hz = 0 ; 1200 Hz = 1). Also the only backward signal it provides is an ON/OFF 387 Hz tone.

600/1200 bit/s Modem CCITT Provisional Rec V.22 (1979)

Nearest equivalents *UK PO modem 27,28*
US –

This is a newly standardised modem which permits full-duplex operation at 1200 bit/s over the PSTN. It employs differential phase modulation using a carrier frequency of 1200 Hz in one direction and 2400 Hz in the other.

Phase modulation is essentially a synchronous technique, but by incorporating a start-stop to synchronous converter, the modem is able to handle start-stop terminals. The modem provides 5 modes of operation, alternative versions having different combinations of these 5 modes.

Mode i) 1200 bit/s synchronous
ii) 1200 bit/s start-stop
8, 9, 10 or 11 bits per character
Alternative A
Alternative B
iii) 600 bit/s synchronous
iv) 600 bit/s start-stop
8, 9, 10 or 11 bits per character
Alternative C
v) asynchronous mode
1200 bit/s start-stop or
300 baud anisochronous

For 1200 bit/s synchronous operation the data stream to be transmitted is divided in dibits, and differential 4-phase modulation is used to code 2 bits per baud. The modulation rate is 600 baud. For 600 bit/s synchronous operation, differential 2-phase modulation is used, coding just 1 bit per baud; the modulation rate remains at 600 baud. Table 4/3 shows the phase changes used. The modem employs a scrambler prior to modulation, to prevent loss of synchronism caused by long strings of repeated dibits. A self-synchronising descrambler is used in the receiving modem to recover the original data after demodulation.

Dibits 1200 bit/s	Bits 600 bit/s	Phase change (modes i–iv)	Phase change (mode v)
00	0	+90°	+270°
01	–	0°	+180°
11	1	+270°	+90°
10	–	+180°	0°

Table 4/3 V.22 Phase shifts

The start-stop to synchronous converter is an elastic buffer in the modem which compensates for any difference between the modem data rate and the start-stop terminal data rate by adding or deleting stop bits as necessary. The modem itself continues to operate in a synchronous fashion, transmitting to line at 1200 or 600 bit/s ± 0.01%. The start-stop terminal may transmit at rates of 1200 or 600 bit/s, + 1%, –2.5%, or optionally with the wider tolerances of + 2.3%, –2.5%. The destination terminal receives data at between 1200 and 1221 bit/s (or 600 and 610 bit/s), with the length of the stop bit being varied by the receiving modem to cope with variations in transmitted data rate.

In mode v) the modem always transmits data at a faster rate than it receives it from the terminal. The permitted terminal rates are:

0-301 bit/s and 1170-1204 bit/s (modem transmits to line at 1205 bit/s)

0-305 bit/s and 1190-1221 bit/s (modem transmits to line at 1223 bit/s)

the range being selected at installation. This mode permits start-stop characters of different lengths to be mixed (providing the lengths only differ by 1 bit, eg lengths of 9 and 10 bits per character).

On PSTN connections, the rule is that the called modem transmits on the high channel, ie:

	Carrier Frequency	
	Transmit	Receive
Calling modem	1200 Hz	2400 Hz
Called modem	2400 Hz	1200 Hz

An 1800 Hz guard tone is also transmitted whenever a modem transmits on the 2400 Hz channel, to prevent misoperation of telephony signalling equipment.

At installation, a V.22C modem is configured to offer mode i) or mode iii) or mode iv) or modes ii) and v). A handshake procedure is specified for PSTN connections to permit automatic selection of either mode ii) or mode v), when this configuration is chosen.

1200/2400 bit/s Modem CCITT Rec V.26 (1968), V.26 *bis (1972)**

Nearest equivalents *UK V.26-PO Modem 7; V.26 bis-PO Modem 12*
US Bell 201

The modem uses differential 4-phase modulation of an 1800 Hz carrier for transmission at 2400 bit/s, and differential 2-phase modulation of the same carrier for transmission at 1200 bit/s. The modulation rate in both cases is 1200 baud.

The V.26 modem provides for full-duplex 2400 bit/s operation over 4-wire private leased circuits. The V.26 *bis* modem provides for half-duplex 2400 bit/s operation over PSTN connections, (or full-duplex over 4-wire leased circuits) with fall-back to 1200 bit/s.

Both modems offer a low-speed (75 bit/s) backward channel using the FSK technique of CCITT Rec V.23.

At 2400 bit/s, two different phase-encoding schemes are defined, as shown in Table 4/4 and Figure 4.1.

Dibit	Phase change	
	Alternative A	Alternative B
00	0°	$+45^\circ$
01	$+90^\circ$	$+135^\circ$
11	$+180^\circ$	$+225^\circ$
10	$+270^\circ$	$+315^\circ$

Table 4/4 V.26 Phase encoding

**bis* – a term used by CCITT to designate the second in a group of related recommendations.

The modem does not incorporate a scrambler, and therefore alternative A is susceptible to loss of synchronism if the data contains long strings of zeros. Alternative B is the one standardised for use on the PSTN. At the fall-back rate of 1200 bit/s, the phase encoding is:

binary 0	$+90^{o}$ phase change
binary 1	$+270^{o}$ phase change

The V.26 *bis* modem designed for operating on the PSTN incorporates a compromise equaliser. Both modems will handle synchronous terminals only.

4800 bit/s Modem CCITT Rec V.27 (1972), V.27 *bis*, V.27 *ter (1976)**

Nearest equivalents *UK* –
US *Bell 208B*

The V.27 and V.27 *bis* modems provide full-duplex 4800 bit/s operation over 4-wire leased circuits. The V.27 *bis* modem offers the additional facility of fall-back to 2400 bit/s. The third member of the family, the V.27 *ter* modem, is designed for use on the switched telephone network; it provides half-duplex transmission at 4800 bit/s with fall-back to 2400 bit/s. All three modems use the same differential phase modulation technique at 4800 bit/s, employing 8 possible phase changes of the carrier, to code 3 bits of data at a time. The carrier frequency is 1800 Hz, and the modulation rate 1600 baud. Table 4/5 and Figure 4.2 show the phase shifts used at 4800 bit/s. A scrambler is incorporated in the modem.

Equalisation

Automatic adaptive equalisation is necessary for operation over the PSTN or over normal-quality private leased circuits. On high-quality private circuits conforming to CCITT Rec M.1020, V.27 specifies only a manual equaliser. This would be satisfactory for point-to-point links, but not for a multipoint network when it would be necessary to go for automatic adaptive equalisation as provided in the V.27 *bis* modem.

Two training sequences for conditioning the receive modem are defined in Rec V.27 *bis* – a long one (708ms) for poor-quality circuits, and a short one (50ms) for high-quality circuits. Longer delays are needed when operating at 2400 bit/s.

Auto-adaptive equalisation is also necessary on the PSTN, and Rec V.27 *ter* defines the training sequences. There is a long sequence for use at initial establishment of the connection, and a short sequence for use when the line is

**ter* – a term used by CCITT to indicate the third in a group of related recommendations.

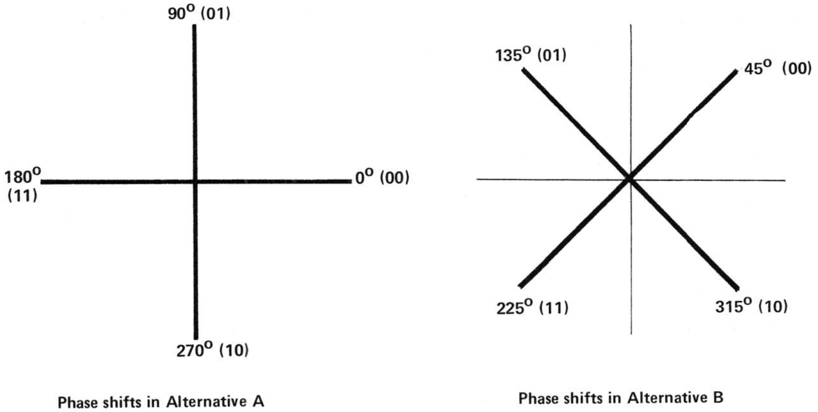

Figure 4.1 Phase Diagrams, Rec V.26

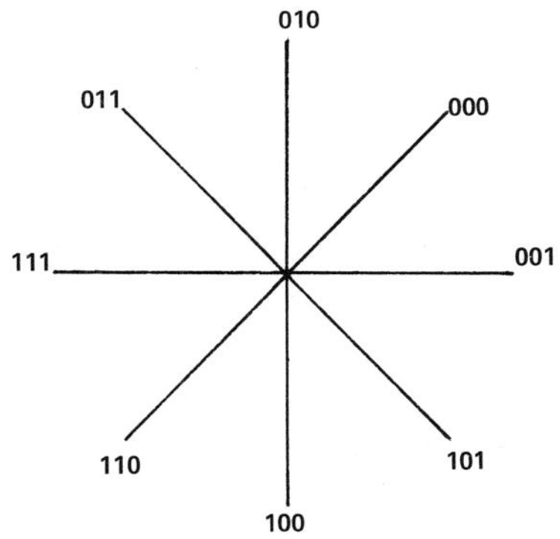

Figure 4.2 Phase Diagram, Rec V.27 ter

'turned round'. The sequences are the same as those used by V.27 *bis* modems, with the exception that there is provision for a longer sequence if protection is required against talker echo.

Fall-back

The 2400 bit/s fall-back facility provided on the V.27 *bis* and V.27 *ter* modems uses 4-phase modulation similar to Rec V.26 alternative A. Table 4/6 shows the phase shifts used.

Backward Channel

A low-speed backward channel of 75 bit/s or higher using the FSK technique of the Rec V.23 backward channel is optionally available on all three modems. This channel can also be used as a secondary data channel to transmit low-speed data in the forward direction.

Tribits			Phase shift
0	0	1	0°
0	0	0	45°
0	1	0	90°
0	1	1	135°
1	1	1	180°
1	1	0	225°
1	0	0	270°
1	0	1	315°

Table 4/5 Phase shifts 4800 bit/s (V.27, V.27 *bis*, V.27 *ter*)

Dibit	Phase shift
00	0°
01	90°
11	180°
10	270°

Table 4/6 Phase shifts 2400 bit/s (V.27 *bis*, V.27 *ter*)

9600 bit/s Modem CCITT Rec V.29 (1976)

Nearest equivalents *UK PO Modem 30*
US –

The standard V.29 modem is designed for full-duplex operation at 9600 bit/s over private leased circuits. Fall-back to 7200 bit/s and 4800 bit/s is possible. The modulation technique used is a mixture of phase and amplitude modulation. The phase modulation part is the same as that employed at 4800 bit/s, and uses 8 possible phase shifts. By permitting the carrier to adopt one of two possible amplitude levels as well, 4 bits per baud can be carried. The modulation rate is thus 9600/4 = 2400 baud. The carrier frequency is 1700 Hz.

At the transmitter, the binary data is divided into groups of 4 ('quadbits'), the last three bits being encoded as a phase shift relative to the previous phase of the carrier according to Table 4/7. Following this shift, the amplitude of the carrier is set to one of two levels, depending on bit 1 (Table 4/8).

Bits 2, 3, 4			Phase Change
0	0	1	0°
0	0	0	45°
0	1	0	90°
0	1	1	135°
1	1	1	180°
1	1	0	225°
1	0	0	270°
1	0	1	315°

Table 4/7 V.29 Phase shifts

Absolute phase	Bit 1	Amplitude
0°, 90°, 180°, 270°	0 1	3 5
45°, 135°, 225°, 315°	0 1	$\sqrt{2}$ $3\sqrt{2}$

Table 4/8 V.29 Amplitudes

Note that different pairs of amplitudes are used depending on the new (absolute) phase of the carrier.

A succinct representation of all the possible states adopted by the modulated waveform is shown in the signal space diagram in Figure 4.3. Each dot represents a possible state.

Fall-Back

Fall-back is possible to 7200 bit/s or 4800 bit/s. At 7200 bit/s, tribit encoding is used, with bit 1 of the quadbit set permanently to zero, giving the signal space diagram shown in Figure 4.4. At 4800 bit/s, the amplitude is held constant, and the modulation becomes pure phase modulation, similar to Rec V.26 alternative A. The modulation rate remains at 2400 baud.

Equalisation

The modem incorporates a self-synchronising scrambler, and an automatic adaptive equaliser, which requires a training period of approximately 253ms.

Multiplexing

The 9600 bit/s V.29 modem is the only CCITT modem to provide multiplexing facilities. The multiplex configurations possible are:

9600; 7200 + 2400; 2 x 4800; 4800 + 2 x 2400; 4 x 2400;
7200; 4800 + 2400; 3 x 2400;
4800; 2 x 2400.

The multiplexed channels may serve either local terminals or remote terminals via another modem link. (See, for example, Figure 5.5, page 91).

48 kbit/s Wideband Modem CCITT Rec V.35 (1968), V. 36 (1976)

Nearest equivalent *UK PO Modem 9 (V.35)*
US –

There are two standard modems which operate at 48 kbit/s. Both are designed to transmit over the 60-108 kHz band which is normally occupied by a group of 12 telephony channels in an FDM telephony system. They are sometimes known as group band modems.

The V.35 modem scrambles the binary data signal and translates it, using a 100 kHz carrier, into the 60-104 kHz band, as an asymmetric sideband suppressed carrier AM signal. It operates at 48 kbit/s synchronous with the option of 40.8 kbit/s synchronous. It can also be used in an anisochronous mode for facsimile transmission where the equivalent bit rate is between 5 bit/s and 48 kbit/s.

V.36 describes a new family of modems to operate full-duplex at 48, 56, 64 and 72 kbit/s over group band circuits. The binary data signal is translated

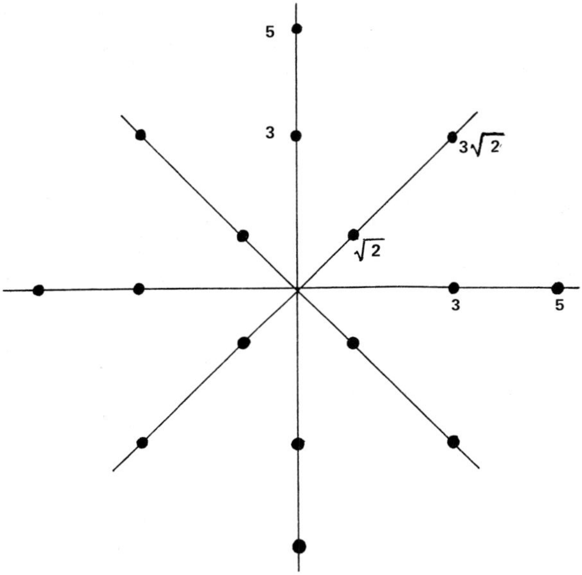

Figure 4.3 Signal Space Diagram Rec V.29, 9600 bit/s

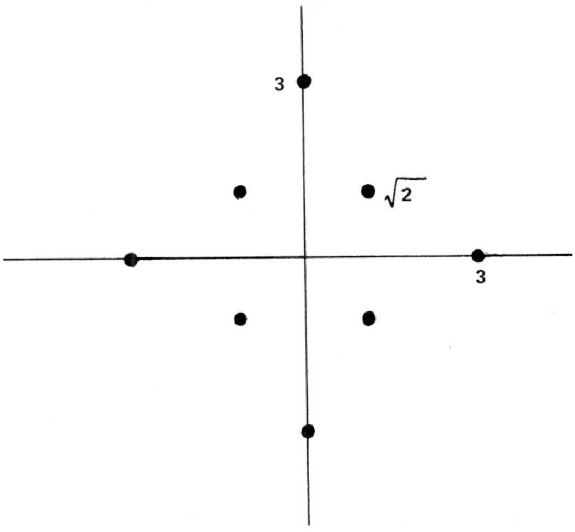

Figure 4.4 Signal Space Diagram Rec V.29, 7200 bit/s

using a 100 kHz carrier into the 60-104 kHz band, as a single-sideband suppressed carrier AM signal.

With both modems, the band between 104 and 108 kHz can be used to provide an optional voice channel.

NON-CCITT MODEMS

In many cases non-standard modems perform at least as well as standard modems. One advantage that standard modems have, however, is that equipment from different manufacturers will interwork, as long as it conforms to the same standard. This is rarely the case for non-standard modems; even if the modulation technique used by different manufacturers is similar, other factors such as synchronisation sequences and scrambler algorithms usually differ and make for incompatibility.

5 Modem Interfaces

When a terminal is connected locally to a computer, the connecting link is completely passive and has no impact on the procedures used by the terminal. When a terminal is connected remotely however, the link – which now includes modems – cannot be disregarded. A modem is an active device, and the terminal has to communicate with the modem as well as with the remote computer. There are thus 2 levels of communication for the terminal, terminal-modem and terminal-computer communication (Figure 5.1).

From the modem's point of view, all the terminal-computer information is classed as 'data'. The modem exists to transport this data along the communications link, the content of the data being irrelevant.

A modem is just one type of Data Circuit-terminating Equipment (DCE), to introduce CCITT terminology. Similarly, Data Terminal Equipment (DTE) is the CCITT term for whatever is connected to the DCE, be it an unintelligent teletypewriter or a mainframe computer. Modems are also known as datasets.

THE DTE-DCE INTERFACE

An interface in this context is a concept, not a piece of hardware. It is a complete description of the boundary between two systems. The boundary between a DTE (terminal) and a DCE (modem) is at the connector by which they are linked, but a complete description of this boundary goes beyond the physical attributes of the connector and includes the logical attributes (the protocol) as well.

The different features of the interface can be listed as follows:

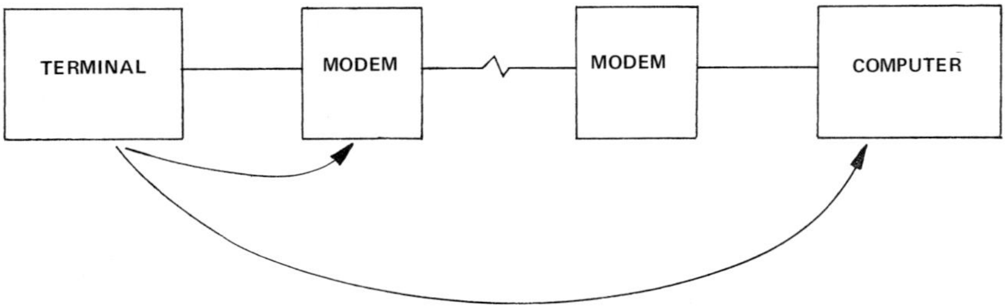

Figure 5.1 Terminal-Modem and Terminal-Computer Communication

Physical Attributes	Logical Attributes
Dimensions and construction of of connector.	Meaning of the electrical signals on each pin.
Number of pins in connector.	Interrelationship between signals.
Electrical signals on the pins.	Procedures for exchanging information between DTE and DCE.

These features have all been standardised internationally, the standards being known generally as V.24 and RS232.

RS232C (the C indicating the current revision) is a recommended standard of the US Electronic Industries Association. The standard is in widespread use in America, and formed the basis of CCITT recommendation V.24, which is common in Europe. RS232C defines all the features listed above. Rec V.24 lists only the DTE-DCE interchange circuits and their functions; the electrical characteristics are defined in another CCITT recommendation, V.28, and the connector pin allocations in an international standard (ISO 2110). However, conformity to these other two standards is usually implied when referring to a 'V.24' interface. For most purposes, V.24 and RS232 can be regarded as synonymous.

(The abbreviations DTE and DCE are similar enough to be confusing, and therefore in the following pages, 'modem' is used instead of DCE, and 'terminal' instead of DTE.)

INTERCHANGE CIRCUITS

In a V.24 interface, signals between the terminal and the modem are carried on separate interchange circuits. One interchange circuit is provided for each function. There are more than 40 interchange circuits in all, which seems excessive at first sight, but then V.24 is a general-purpose interface covering a wide range of modem applications, and no single modem would use all the interchange circuits.

There are in fact 2 sets of interchange circuits, the 100-series used for data, timing and control circuits, and the 200-series used for automatic calling. Of the 100-series circuits, there is a core of 8 circuits which is common to many applications, and these are listed in Table 5/1.

Circuit V.24 (RS232C)	Designation	Direction	
		To Modem	To Terminal
102 (AB)	Signal ground or common return		
103 (BA)	Transmitted data	X	
104 (BB)	Received data		X
105 (CA)	Request to send	X	
106 (CB)	Ready for sending (Clear to send)		X
107 (CC)	Data set ready		X
108/1 –	Connect data set to line	X	
or			
108/2 (CD)	Data terminal ready	X	
109 (CF)	Data channel received line signal (ie carrier) detector		X

Table 5/1 Main V.24 Interchange Circuits

The operation of these circuits is best understood by considering the sequence of events associated with data transmission over the telephone network. This is a simplified description aimed at giving an overall view of the procedure; a detailed description of the operation of each interchange circuit is given later.

When a data connection is required over the PSTN, the first step is to dial the required number on the telephone in the usual way. When the call is answered and a communication path exists between the two parties, the telephone line needs to be switched at each end from the telephone to the data terminal. This is done by the modem once the data terminal has turned ON circuit 108. (There are two slightly different ways of using this circuit as described later. The circuit is designated Connect Data Set to Line (108/1) or Data Terminal Ready (108/2) to distinguish the two.)

When connected to line, the modem informs the terminal by turning ON the Data Set Ready circuit 107 (CC). If the distant modem is already connected to line and is transmitting, the local modem will turn ON the Carrier Detect circuit 109 (CF) to indicate that it is detecting a carrier signal. This is often indicated by a light on the terminal. Any data that is received will be passed to the terminal over the Receive Data circuit 104 (BB). Figure 5.2 shows this sequence of events.

If a terminal wishes to transmit, it turns ON the Request to Send circuit 105 (CA). When the modem is ready to accept data for transmission, it replies by turning ON the Clear to Send circuit 106 (CB). Data can then be transmitted by the terminal on the Transmit Data circuit 103 (BA).

The one circuit not mentioned so far is the Signal Ground circuit 102 (AB). This provides the essential common return lead for all the other interchange circuits. In the future, equipment may use two common return circuits, one from the terminal to the modem, and the other from the modem to the terminal (see p 105).

MAJOR CONTROL CIRCUITS

Data Channel Received Line Signal (ie Carrier) Detector Circuit 109 (CF)

Modem to Terminal

The Carrier Detect circuit 109 (CF) is turned ON by the modem to indicate to the terminal that a carrier signal which meets the necessary criteria for signal level, duration etc is being received from line. There is a delay between a carrier arriving at the modem and Carrier Detect turning ON to allow time for the modem to synchronise. This delay will be at least 5-20ms. Longer delays are needed when the modem employs automatic equalisation, as Table 5/2 indicates.

When automatic calling and answering are implemented, there may be a delay of 300-700ms at call establishment only.

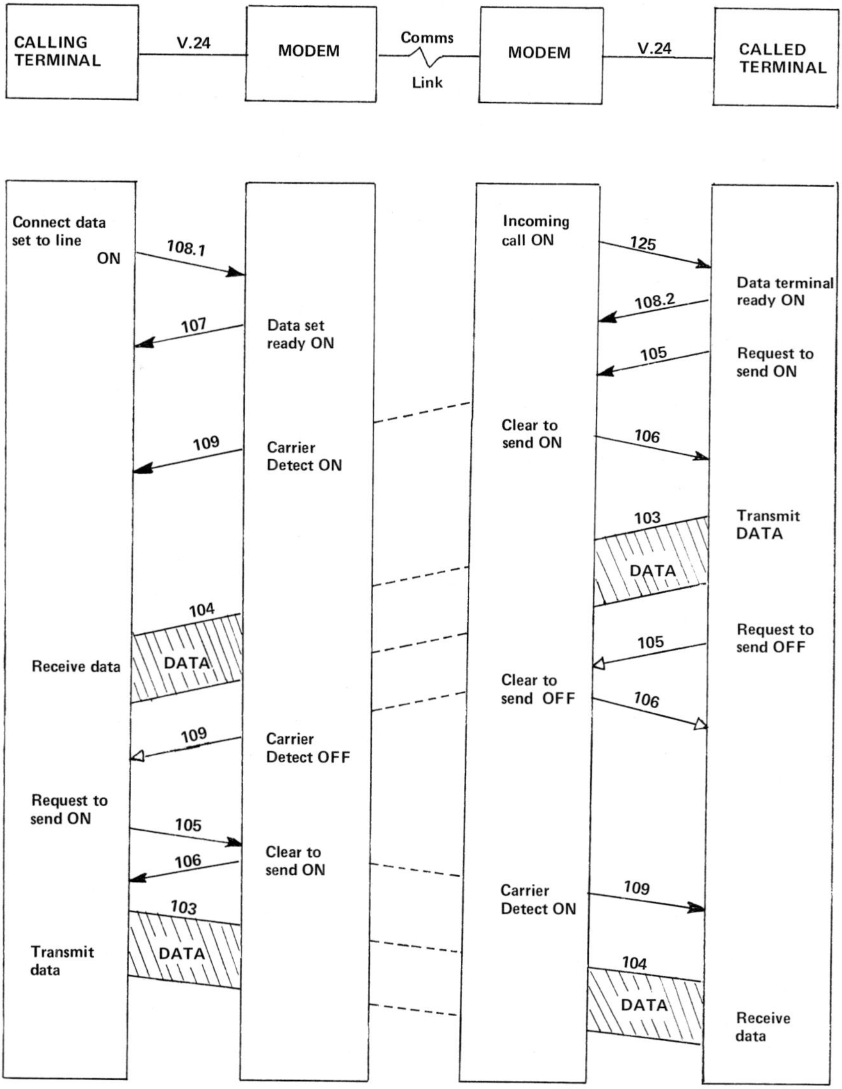

Figure 5.2 V.24 Interface Procedure

Modem	Delay (milliseconds)	
	PSTN	Leased circuit
4800 bit/s V.27 *bis*	–	50 or 708 (note 1)
V.27 *ter*	708 at call establishment (note 2) 50 at turn-round (note 2)	–
9600 bit/s V.29	–	Approx 253 initially. 5-15 subsequently if continuous carrier mode is used

Note 1. The longer times may be needed on poor quality circuits.

Note 2. These times should be increased by 215ms if the modem transmits unmodulated carrier prior to the training sequence, in order to protect against talker echo.

Table 5/2 Turn-on Times of Circuit 109 (CF) Carrier Detect

Turn-off delays on circuit 109 (CF) are normally 5-15ms. Standard 200 baud V.21 modems use a longer turn-off delay of 20-80ms.

Circuit 122 (SCF) is the Carrier detect interchange circuit for the low-speed backward channel. When implemented the backward channel employs the FSK technique of Rec V.23, and the appropriate delays are:

turn-on < 80ms

turn-off 15-80ms

Request to Send RTS Circuit 105 (CA) *Terminal to Modem*
Ready for Sending/Clear to Send Circuit 106 (CB) *Modem to Terminal*

A terminal turns ON the Request to Send circuit 105 (CA) when it wishes to transmit data. Unless the modem is of the continuous carrier variety, this causes a carrier signal to be transmitted. Data transmission can begin as soon as the modem turns ON the Clear to Send circuit 106 (CB) in response. This response may be delayed by a preset amount by the modem.

Modems usually provide a number of different delays; the one chosen depends on the application and the data network configuration.

For minimum delay, continuous carrier operation should be employed. If the modem transmits carrier continuously, there is no need to resynchronise the modems every time the flow of data is reversed. However, continuous carrier operation is only possible on full-duplex circuits.

When the carrier is switched on and off under control of RTS, the delay before CTS comes ON must be long enough for the receiving modem to attain synchronism, and if necessary, equalise to the line conditions. This delay has already been discussed in relation to the Carrier Detect circuit 109 (CF). In multipoint networks, it is common to find the central modem operating in continuous carrier mode, and the outstation modems in controlled carrier mode.

In half-duplex 2-wire operation, a turn-on delay of about 200ms is needed when 'turning-round' (reversing the direction of transmission) to allow the line to settle down and echoes to die away. (To avoid this delay, which is wasted time, many applications use full-duplex 4-wire circuits even though the data flow is essentially half-duplex.)

On PSTN connections, the use of automatic calling and answering equipment may introduce a turn-on delay of several seconds when the connection is first established.

No delay is introduced between RTS being turned OFF and CTS going OFF.

When a low rate backward channel is provided, circuits 120 (SCA) and 121 (SCB) provide the functions of Request to Send/Clear to Send for the backward channel. The delay between the terminal turning ON (S)RTS (circuit 120-SCA) and the modem responding by turning on ON (S)CTS (circuit 121-SCB), is 80-160ms in accordance with Rec V.23.

Calling (Ringing) Indicator Circuit 125 (CE) *Modem to Terminal*

This circuit is used for automatic answering and alerts the terminal to an incoming call. The circuit reacts in sympathy with the ringing current, turning ON during 'rings'. Figure 5.3 indicates the UK ringing tone.

Circuit 125 is independent of the other interchange circuits, and remains operational during modem testing.

Connect Data Set to Line (CDSTL) Circuit 108/1 *Terminal to Modem*

This circuit gives the terminal *direct* control over switching the modem to the telephone line, for connections set up over the PSTN. The call is dialled with circuit 108/1 OFF, and when the call is answered this circuit is turned ON by the terminal to disconnect the telephone and connect the modem to line.

The circuit is an alternative to DTR, circuit 108/2. Whichever circuit is implemented the modem responds via the Data Set Ready circuit 107 (CC).

Data Terminal Ready (DTR) Circuit 108/2 (CD) *Terminal to Modem*

This circuit provides the terminal with *indirect* control over switching the modem to line; it indicates to the modem that the terminal is ready for the modem to be switched to line. The actual switching is accomplished by some other means, such as a push-button on the telephone in the case of manually originated or answered calls.

DTR is provided as an alternative to CDSTL (108/1), and is the circuit implemented when the modem has automatic answering facilities. A terminal ready to receive incoming calls maintains DTR ON, so that when a call is received and the Calling Indicator circuit 125 is turned ON, the modem is automatically switched to line. This happens at the end of the first cycle of ringing tone, and is indicated to the terminal by the Data Set Ready circuit 107 coming ON. DTR cannot be permanently strapped ON, because the terminal needs to control it in order to clear down PSTN calls.

DATA CIRCUITS

Transmitted Data Circuit 103 (BA) *Terminal to Modem*

This is the interchange circuit over which the terminal transmits data. The modem will only accept data transmitted over this circuit if the four interchange circuits RTS (105), CTS (106), DSR (107) and CDSTL (108/1)/DTR (108/2) are all ON; this condition is the result of the handshaking procedure shown in Figure 5.4.

(If circuit CDSTL/DTR or RTS is not implemented in a particular application, the modem will need to be strapped to see an ON condition on the missing circuit.)

Transmission of data continues until one of these circuits is turned OFF. Normally the terminal would switch OFF either RTS to signify that it had finished transmitting and was ready to receive, or CDSTL/DTR to signify the end of the call.

When not transmitting, a terminal usually maintains a binary 1 condition on the Transmitted data circuit; the same condition may also be used to fill gaps between characters during data transmission ('idle mark').

Received Data Circuit 104 (BB) *Modem to Terminal*

This is the interchange circuit over which the terminal receives data. The circuit is subject to unwanted signals such as switching transients, echoes, and other forms of noise on the communications link, and to protect the terminal

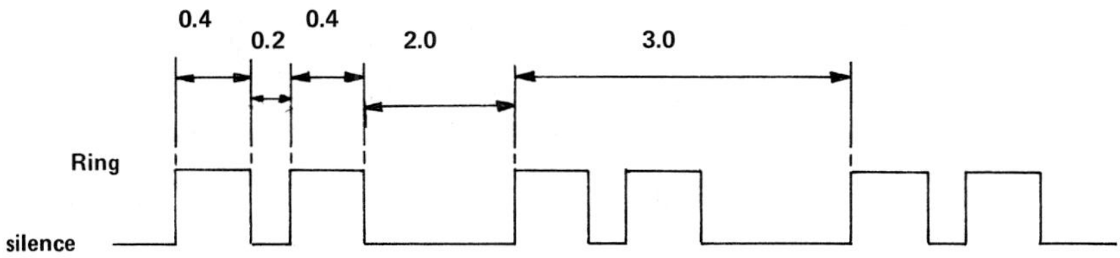

Figure 5.3 UK Ringing Tone (times in seconds)

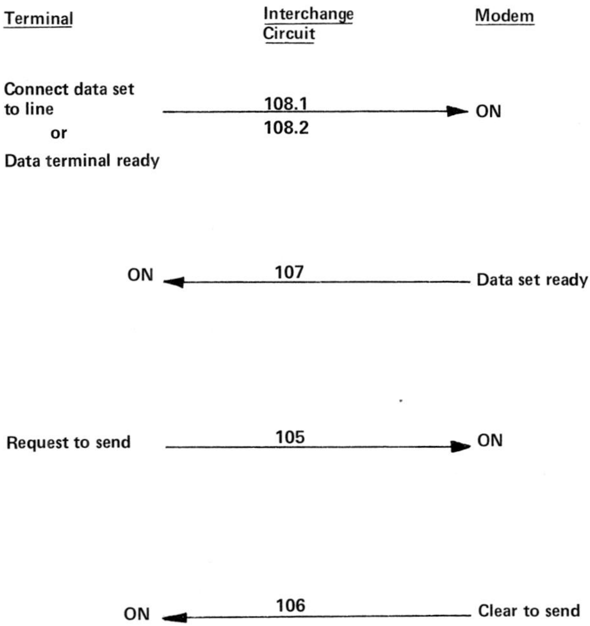

Figure 5.4 V.24 Initial Handshake

from these spurious signals the modem may clamp the circuit to binary 1 whenever the Carrier Detect circuit 109 (CF) is OFF.

Circuit 104 (BB) may also be clamped to binary 1 in half-duplex operation whenever the terminal is transmitting, ie whenever RTS is ON. This prevents the transmitted signal being fed back to the terminal. Removal of this clamp may be delayed for up to 175ms after RTS is turned OFF, to allow for completion of transmission and to protect the terminal from false signals such as synchronisation sequences.

TIMING CIRCUITS

In synchronous operation the data to be transmitted is clocked into the modem from the terminal at a steady rate. The clock which provides this timing may be located within the modem or may be external to the modem, in the terminal. The modulated waveform transmitted by the modem contains timing information which allows the destination modem to clock out the data to its terminal at the same steady rate.

There are four V.24 interchange circuits for conveying clock signals between modem and terminal.

Circuit 113 (DA) Transmitter signal element timing – DTE source (to modem, clock in terminal)

Circuit 114 (DB) Transmitter signal element timing – DCE source (to terminal, clock in modem)

These two circuits are used to time the data sent to the modem on the Transmit Data circuit 103 (BA). Either terminal timing (circuit 113-DA) or modem timing (circuit 114-DB) would be provided but not both. Usually data is transmitted to the modem under control of the modem clock, using the Transmit clock timing circuit 114. External transmit timing is used in applications like the one shown in Figure 5.5 where modems are connected back-to-back, and timing is derived from one single source.

Circuit 115 (DD) Receiver signal element timing – DCE source (to terminal, clock in modem)

Circuit 128 Receiver signal element timing – DTE source (to modem, clock in terminal)

These two circuits are used to time the received data on circuit 104 (BB). The Receive clock signal on circuit 115 tells the terminal when to sample the received data on circuit 104. This timing is derived from the incoming modulated signal and is therefore synchronised to the timing used at the transmitter (Figure 5.6).

Circuit 128 is rarely implemented. It enables the terminal to clock in the

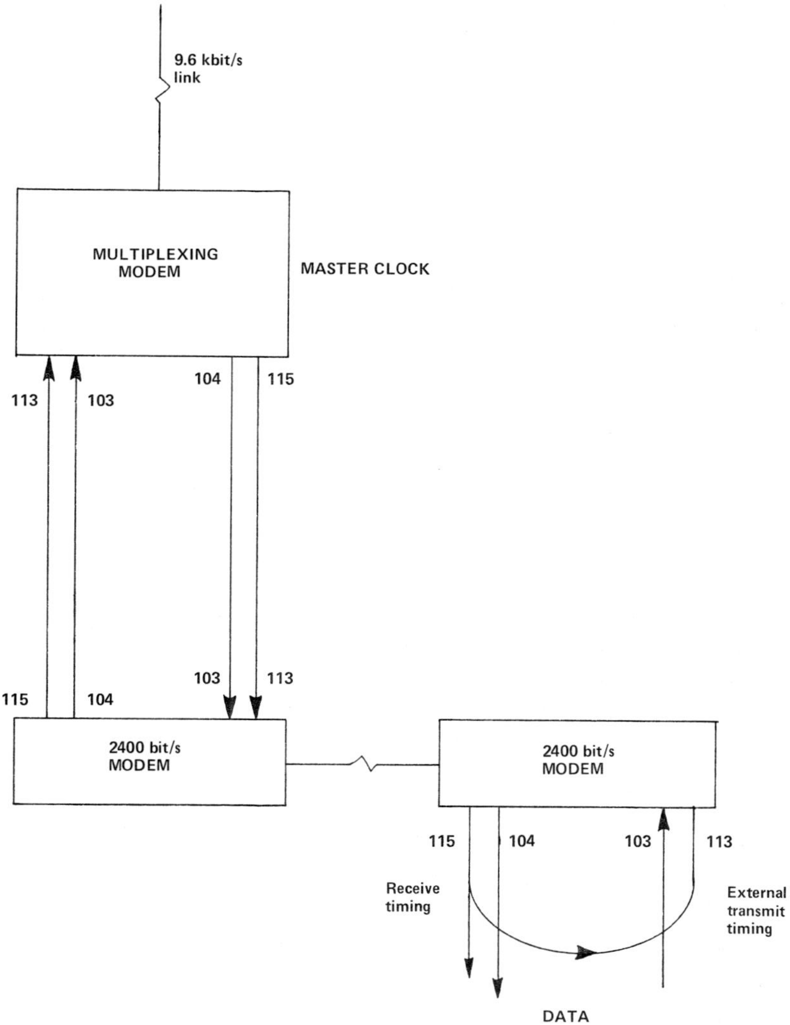

Figure 5.5 A Remote Tail Circuit Synchronised to the Clock in the High-Speed Multiplexing Modem

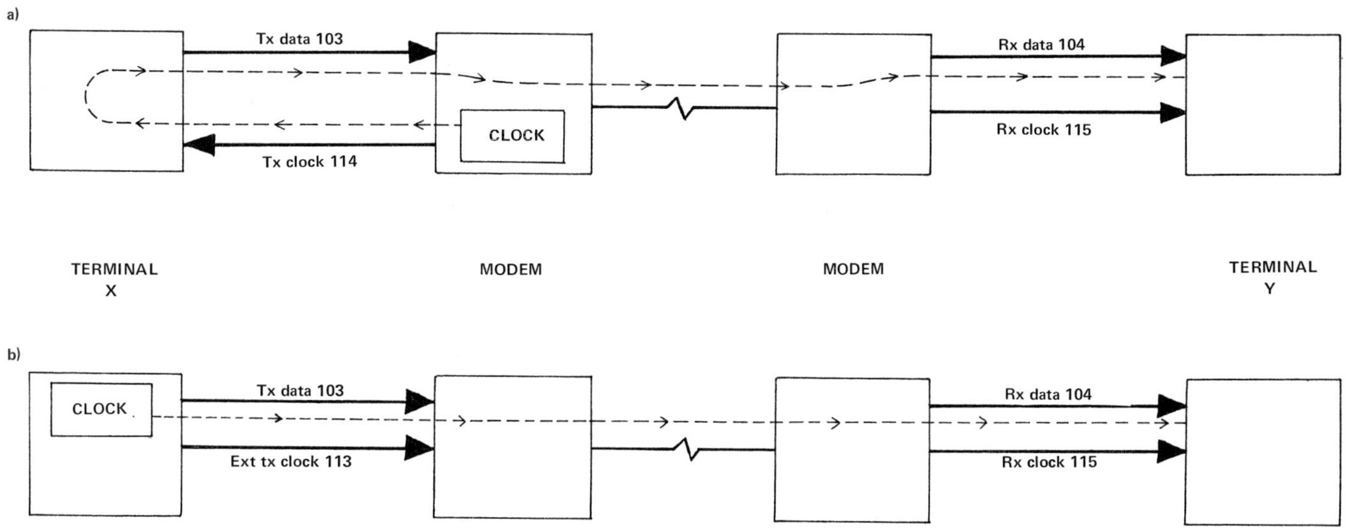

Figure 5.6 Distribution of Timing Signals X to Y (– – – –)
a) Clock in Modem b) Clock in Terminal

received data on circuit 104 in its own time. This circuit could be used in conjunction with a synchronous modem which had asynchronous standby facilities.

On the timing circuits, the ON to OFF transition nominally coincides with the centre of the data bits on the transmit or receive circuits, and the OFF to ON transition with transitions in the data signals (Figure 5.7).

Timing from the modem clock is normally provided to the terminal whenever the modem is powered up, although signals may be suspended for short intervals during modem testing. Receive data clock on circuit 115 may not be available when the Carrier Detect circuit 109 is OFF.

OTHER CONTROL CIRCUITS

Data Signal Quality Detector Circuit 110 (CG) *Modem to Terminal*

Not commonly implemented, this circuit is designed to indicate to the terminal that the received data may be corrupted. Some modems now provide an indicator light which performs the same function. An ON condition indicates 'good' data.

Data Signalling Rate Selector (DTE) Circuit 111 (CH) *Terminal to Modem*

This circuit is implemented on dual rate modems such as the standard 4800/2400 bit/s V.27 *ter* modem. Circuit 111 (CH) allows the terminal to select the data rate required; an ON condition selects the higher rate or range of rates, and an OFF condition the lower rate or range of rates.

Data Signalling Rate Selector (DCE) Circuit 112 (CI) *Modem to Terminal*

This rarely implemented circuit is used to select one of two data signalling rates or ranges of rates in the terminal to coincide with those used in the modem. The ON condition selects the higher rate or range, and the OFF condition the lower rate or range.

Select Standby Circuit 116 *Terminal to Modem*
Standby Indicator Circuit 117 *Modem to Terminal*

Circuit 116 allows the terminal to select standby facilities such as a lower transmission rate or alternative communications links. For example, some modems operating at 2400 bit/s over leased lines offer fall-back operation at 1200/600 bit/s over the PSTN. Circuit 116 is used to select normal or fall-back operation, and for the latter, circuit 111 would be used to determine whether to operate at 600 or 1200 bit/s.

Circuit 117 indicates to the terminal whether normal or standby facilities have been selected. For both circuits, the ON condition is used for standby,

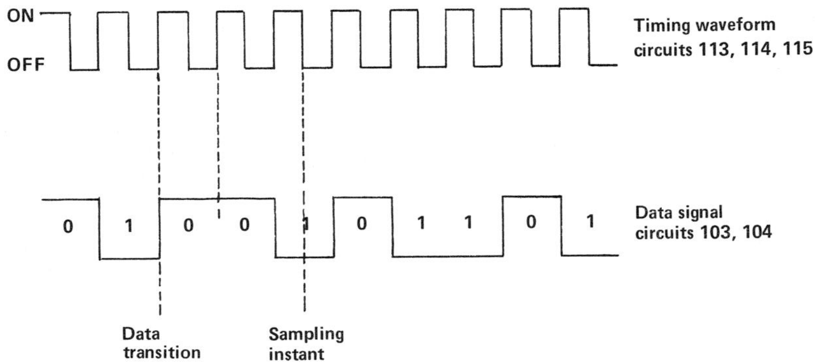

Figure 5.7 Relationship of Timing and Data Circuits

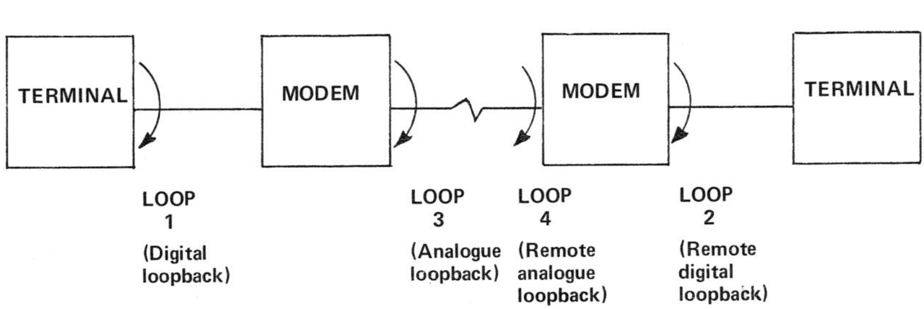

Figure 5.8 Loopback Points

and the OFF condition for normal operation.

Select Transmit Frequency Circuit 126 *Terminal to Modem*
Select Receive Frequency Circuit 127 *Terminal to Modem*

These circuits were designed for the standard 200/300 baud V.21 modem, which uses different transmit frequencies for the two directions of transmission. Usually on PSTN connections, however, the choice of frequencies is made automatically by the modem, depending on whether it is the called or calling party, and these circuits are not required. On some multipoint applications control of the modem frequencies by the terminal may be required, but often circuit 126 will be used to control both transmit and receive frequencies and circuit 127 will not be implemented. For both circuits, an ON condition signifies the higher and the OFF condition the lower frequency.

SECONDARY BACKWARD CHANNEL CIRCUITS

The low-speed backward channel operating at 75 bit/s (or up to 150 bit/s in some modems) has a complete set of data and control circuits matching those of the main forward channel, as shown in Table 5/2.

Circuit V.24 (RS232)	Designation	Direction		Main Channel equivalent
		To Modem	To Terminal	
118 (SBA)	Transmitted backward channel data	X		103 (BA)
119 (SBB)	Received backward channel data		X	104 (BB)
120 (SCA)	Transmit backward channel line signal	X		105 (CA)
121 (SCB)	Backward channel ready		X	106 (CB)
122 (SCF)	Backward channel received line signal detector		X	109 (CF)
123	Backward channel signal quality detector		X	110 (CG)

Table 5/2 Backward Channel Interchange Circuits

The backward channel employs asynchronous frequency shift keying, and no timing circuits are necessary.

INTERCHANGE CIRCUITS FOR MODEM TESTING AND FAULT ISOLATION

A loopback test, in which the transmitted data is looped back so that it appears on the received data path, is an effective means of fault isolation. On a simple point-to-point connection, there are four locations at which it may be suitable to loopback the circuit. These are shown in Figure 5.8.

Loop 1 checks the terminal, loop 3 the local modem, and loop 2 the remote modem. These loops may be activated by a switch on the modem, or via a terminal-modem interchange circuit. Loop 4 (which is only possible on 4-wire circuits) is for PTT use only. The circuits provided in V.24 are:

Terminal to Modem:

Remote loopback circuit 140

– activates loop 2 in Figure 5.8

Local loopback circuit 141

– activates loop 3 in Figure 5.8

Modem to Terminal:

Test indicator circuit 142

The modem switches this circuit ON in response to a loopback command from the terminal on circuit 140 or 141. Circuit 142 is also switched ON when the modem is tested from a remote location. Data transmission is impossible when circuit 142 is ON.

As data networks have grown in complexity, the need for adequate monitoring and diagnostic aids has grown in importance. Many modem manufacturers now provide centralised network control equipment, which is able to communicate with all the modems on a network. With such equipment it becomes possible to monitor the V.24 interface circuits at each modem, perform loopback tests, switch in standby modems, change over to spare circuits or alter data transmission rates – all remotely. Communication can take place over the modem's low-speed secondary or backward channel, or over a separate dialled connection, and so can occur while data is being transmitted on the main channel.

200-SERIES CIRCUITS FOR AUTOMATIC CALLING

An automatic calling unit allows calls to be set up over the PSTN without manual intervention. The unit is interposed between the terminal and the modem and is connected to both (Figure 5.9).

The terminal passes the digits to be dialled to the automatic calling unit, and the unit converts these to dial pulses (or to multifrequency tones for

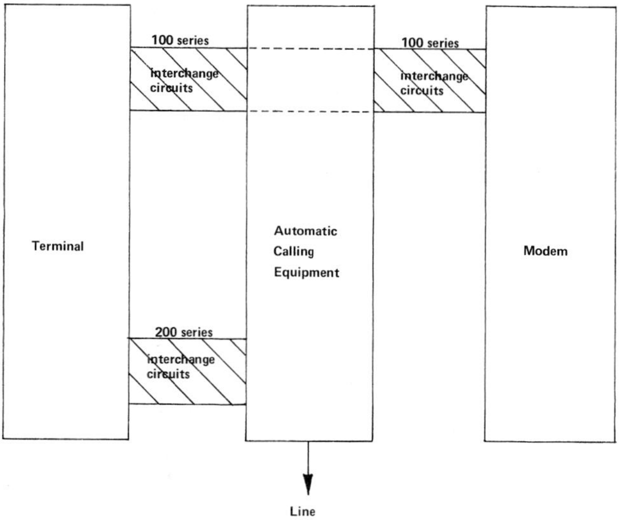

Figure 5.9 Connection of Automatic Calling Equipment

telephone networks so equipped). Having sent all the digits, the unit causes the modem to transmit a calling signal to line to announce the fact that the call is being originated automatically. The calling signal comprises short bursts of 1300 Hz or other binary 1 tone repeated every 1½–2 seconds.

The automatic calling unit relies on detecting a 2100 Hz answering tone before it will connect the calling terminal to line. If no such tone is received within a specified period the unit advises the terminal to abandon the call.

There are twelve interchange circuits in the 200-series: four for data, seven for control and indication, and the common return circuit. They are listed in Table 5/3.

The procedure for making a call is illustrated in Figure 5.10 which is based on the CCITT recommendation V.25. The terminal turns ON the Call Request circuit 202 causing the automatic calling unit to go 'off-hook', which is indicated by the Data Line Occupied circuit 203 coming ON. When dial tone is received the unit invites the terminal to 'Present next digit' via circuit 210. The terminal presents the first digit to be dialled in parallel form on the 4 circuits 206-209, coded as shown in Table 5/4, and informs the unit by turning ON the Digit Present circuit 211. After dialling the first digit, the unit turns 210 OFF. The terminal responds by turning 211 OFF, and then the unit turns 210 ON to request the second digit. The procedure is repeated until all the digits have been sent, which is signified by the terminal placing

Circuit	Designation	Direction: To terminal	Direction: From terminal
201	Common return		
202	Call request		X
203	Data line occupied	X	
204	Distant station connected	X	
205	Abandon call	X	
206	Digit signal 2^0		X
207	Digit signal 2^1		X
208	Digit signal 2^2		X
209	Digit signal 2^3		X
210	Present next digit	X	
211	Digit present		X
213	Power indication	X	

Table 5/3 200-Series Interchange Circuits in V.24 for Automatic Calling

the End of Number code on circuits 206-209. Delays between digits, to allow for a second dial tone for example, can be introduced by presenting the Separation Control Character SEP between digits.

The calling tone is transmitted for a preset time between 10 and 40 seconds, and in the absence of a 2100 Hz answering tone the unit turns ON the Abandon Call circuit 205, to which the DTE must respond by turning 202 OFF.

If the call is answered, the unit detects a 2100 Hz tone from the distant modem. This is allowed to persist for 450-600ms to ensure that any echo suppressors have been disabled, after which the unit transfers control of the line from circuit 202 to DTR (circuit 108/2). The 2100 Hz tone lasts for 2.6–4.0 seconds, and about 75ms after it ceases, DSR (circuit 107) is turned ON indicating that unit has completed its calling procedures. Receipt of carrier from the distant modem causes the Distant Station Connected circuit 204 to be turned ON, after which the automatic calling unit plays no further part. Circuit 202 can then be turned OFF by the terminal and control of the connection is vested in DTR (circuit 108/2), as for a normal call.

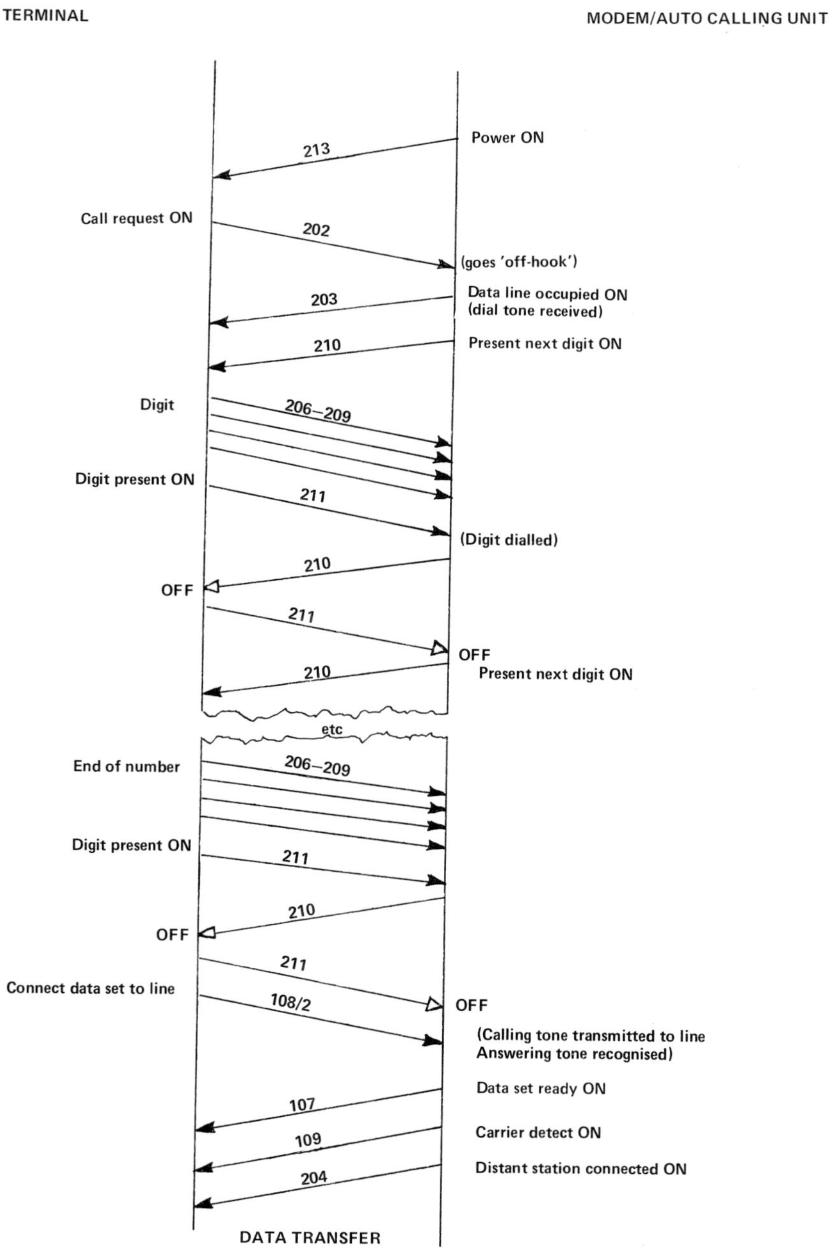

Figure 5.10 V.25 Call Set-up Procedure

The US standard for automatic calling is RS366, and it uses the same circuits as CCITT recommendation V.25. The circuit designations differ slightly (see Table 7/3, page 117).

Recommendation V.25 also defines the procedures for automatic answering. Many modems have automatic answering facilities today, and for those that do not, the PTTs can supply automatic answering devices to interpose between the terminal and the modem.

When an incoming call is received, an auto-answer modem signals to the terminal over the Calling circuit 125, and assuming DTR is ON, goes 'off-hook' on completion of the first ringing cycle. The modem waits for 1.8–2.5 seconds, and then transmits 2100 Hz tone for 2.6–4.0 seconds. Like the modem at the calling end, the called modem turns on the Data Set Ready circuit 107 about 75ms after the 2100 Hz tone ceases.

Signal	Circuit state			
	209	208	207	206
digit 1	0	0	0	1
2	0	0	1	0
3	0	0	1	1
4	0	1	0	0
5	0	1	0	1
6	0	1	1	0
7	0	1	1	1
8	1	0	0	0
9	1	0	0	1
0	0	0	0	0
End of Number	1	1	0	0
Separation Control Character	1	1	0	1

Table 5/4 Code Combinations, Circuits 206–209

6 Electrical Characteristics of the V.24 Interface

Signalling at the V.24 interface is effected by means of different voltage levels on the interchange circuits. A positive voltage represents a 0 or ON condition, and a negative voltage a 1 or OFF condition. The magnitude of the voltage varies, but $\pm$ 12 volts is common.

The complete electrical specification for the traditional V.24 interface is given in CCITT recommendation V.28, which is virtually identical to the electrical specification contained in RS232C. These two standards apply to data rates up to 20 kbit/s, and at one time the only standard for operation above this rate was contained in CCITT recommendation V.35, which relates to 48 kbit/s modems. Now, however, there are two new standards for electrical characteristics at the interface, known variously as V.10/X.26/RS423 and V.11/X.27/RS422, and these provide for operation up to 100 kbit/s and 10 Mbit/s respectively.

These two standards are only beginning to make an impact, but in the long term they are expected to replace the existing ones. Old and new standards are discussed below.

CCITT REC V.28 AND RS232C

Rec V.28 is entitled *Electrical characteristics for unbalanced double current interchange circuits*, and is identical to the corresponding parts of RS232C, with one minor exception which will be discussed.

We start with definitions: an *interchange circuit* has two leads to provide a 'go' and 'return' path for the current; a *double-current* interchange circuit is one on which current may flow in either direction, depending on the polarity of the voltage applied; an *unbalanced* interchange circuit is one in which one of the leads is at ground (earth) potential.

A complete interchange circuit can be regarded as a voltage generator connected to a load, as in Figure 6.1. In the case of the Transmitted Data circuit 103, for example, the generator would be in the terminal and the load in the modem. The line of demarcation between the generator and the load is at the V.24 connector attached to the modem.

When disconnected, the voltage produced by the generator must not exceed 25 volts. When connected to a load having a resistance anywhere between 3000 and 7000 ohms, the voltage must remain between 5 and 15 volts. The capacitance of the load, including the cable, must not exceed 2500pF.

The threshold voltages are defined as $\pm$ 3 volts (Table 6/1), giving a 2-volt margin between the threshold and the minimum permitted voltage of 5 volts.

V less than −3 volts	V greater than +3 volts
1	0
OFF	ON

Table 6/1 V.28 Correlation

The region between −3 and +3 volts is known as the transition region. When a circuit is switched ON or OFF, the voltage should pass smoothly through this transition region, as in Figure 6.2, at a rate of less than 30 volts/microsecond to minimise crosstalk. For data and timing circuits, the time taken to cross the transition region, 't' in Figure 6.2, is specified as follows:

V.28: t = 1ms or 3% of the bit period, whichever is less.

RS232C: t = 1ms or 4% of the bit period, whichever is less.

The time t depends on the capacitance of the load and thus is affected by the length of the interconnecting cable. Other things being equal, the wider tolerance in RS232C would permit slightly longer interchange circuits than V.24.

Distance Limitations

Modem-terminal interchange cable typically has a capacitance per circuit of about 150 picofarads/metre. This limits the maximum length to about 16 metres if the maximum load capacitance of 2500pF specified in the standard is not to be exceeded. RS232C actually sets a maximum interchange circuit length of 50 feet (15.3 metres). V.28 sets no such limit, (although the UK Post Office suggests 30 metres up to 2400 bit/s).

In practice much longer distances can be traversed; operation over half a mile at 1200 bit/s is not impossible. It all depends on the environment and the cable used, and on whether transmission is synchronous or asynchronous. For long distances the requirements are:

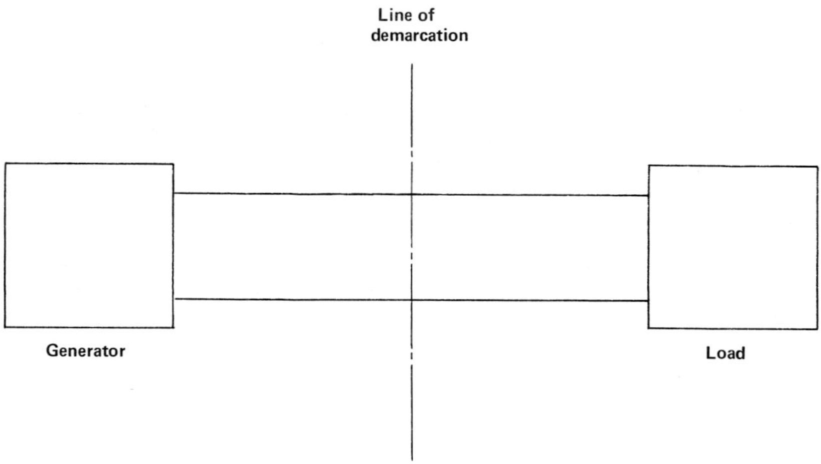

Figure 6.1 V.28 Interchange Circuit

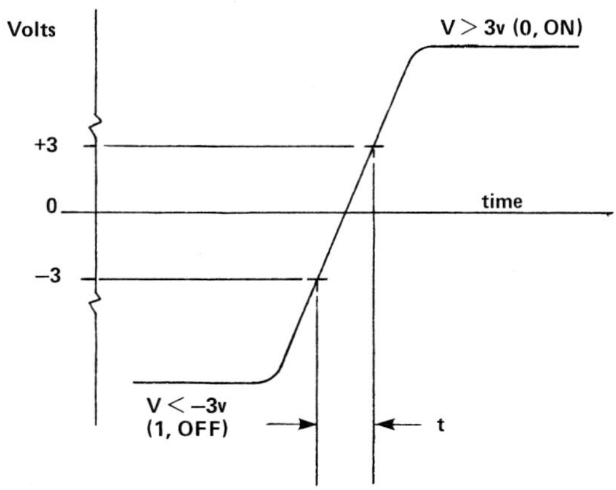

Figure 6.2 V.28 Electrical Characteristics

- an electrically 'quiet' environment, to minimise induced currents in the interchange cable which can lead to signals being misinterpreted;
- low capacitance cable, to minimise pulse degradation;
- asynchronous transmission, to avoid crosstalk caused by clock pulses on adjacent circuits.

McNamara* gives a good practical treatment of this subject with results of data rate versus distance experiments.

In some circumstances line drivers may offer a solution. These amplify the signals on the data and timing circuits and, operating usually over twisted pairs of wires, can extend the range of the interface considerably. They do, however, increase the risk of causing interference to other circuits in the vicinity.

CCITT REC V.35 ELECTRICAL CHARACTERISTICS

The electrical characteristics specified in V.35 are for balanced interchange circuits operating at 48 kbit/s. Balanced circuits have a lower capacitance per unit length than unbalanced circuits, and therefore degrade high-frequency signals less. A balanced circuit comprises a twisted pair of wires, the two leads – known as the A and B wires – carrying equal and opposite voltages. Unlike an unbalanced circuit, neither lead is at earth potential. This provides a cancellation effect, so that a balanced circuit is less prone to interference from outside and also less likely to cause interference than an unbalanced circuit.

The voltage used in V.35 is 0.55v ± 20%, for an interchange circuit terminated in a 100 ohm load. Table 6/2 shows the correlation between voltages and binary signals.

$V_A - V_B = -0.55v$	$V_A - V_B = +0.55v$
1	0

Table 6/2 V.35 Correlation

No distance limitation for the interchange circuits is laid down in V.35. The UK Post Office suggests 60 metres.

* *Technical Aspects of Data Communication*, John E McNamara, Digital Equipment Corporation.

NEW ELECTRICAL CHARACTERISTICS

The V.28/RS232 standard was developed in the days of discrete electronic components, and the electrical characteristics defined therein are not compatible with today's integrated circuit technology. The distance limitation of V.28/RS232 has also presented problems in the past.

The two new standards already referred to are an improvement in many ways, in that they are IC compatible, permit operation over much longer distances, and operate at much higher data rates. They originated as recommended standards of the Electronic Industries Association (EIA) in America, where they are known as RS423 and RS422. They were subsequently adopted by CCITT, where they were designated recommendations V.10 and V.11 respectively. However, CCITT also intends to use these new electrical characteristics on interfaces to public data networks, and has assigned them two other numbers, X.26 and X.27. Thus we have:

CCITT Rec V.10 = CCITT Rec X.26 = EIA RS423
and CCITT Rec V.11 = CCITT Rec X.27 = EIA RS422.

To compound the confusion, EIA have also issued a standard RS449, which covers the non-electrical aspects of the interface (such as connectors) and updates RS232C. US equipment employing the new electrical characteristics is sometimes referred to as RS499 equipment. For brevity in the following sections, the two standards will be referred to merely as V.10 and V.11.

V.10 is designed for unbalanced interchange circuits, and V.11 for balanced interchange circuits. The lower capacitance per unit length of a balanced interchange circuit means that it can be used over longer distances and at higher data rates than an unbalanced circuit.

CCITT Rec V.10

V.10 operates up to 100 kbit/s. It uses voltages between ± 3 and ± 6 volts or, if limited interworking with V.28 is required, ± 4 and ± 6 volts. In both cases, the threshold voltage is only 0.3 volts.

The equivalent circuit for a V.10 interchange circuit is shown in Figure 6.3. The receiver is a differential receiver, with two inputs; it measures the voltage between these two inputs, not the voltage relative to ground. Differential receivers are specified for V.10 for compatibility with V.11, so that balanced and unbalanced circuits can be mixed in the same interface if necessary. In unbalanced operation, however, one of these inputs would in fact be grounded, and connected to the common return circuit. Separate common return circuits, designated 102a and 102b, are used for each direction.

Table 6/3 shows the correspondence between electrical and logical conditions.

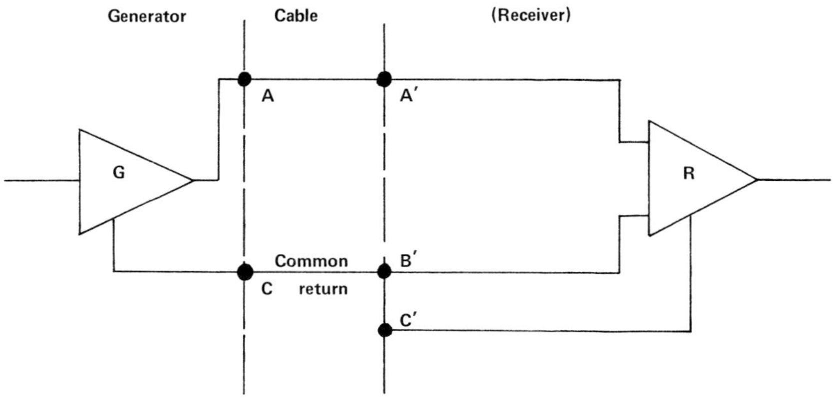

Figure 6.3 V.10 Equivalent Circuit

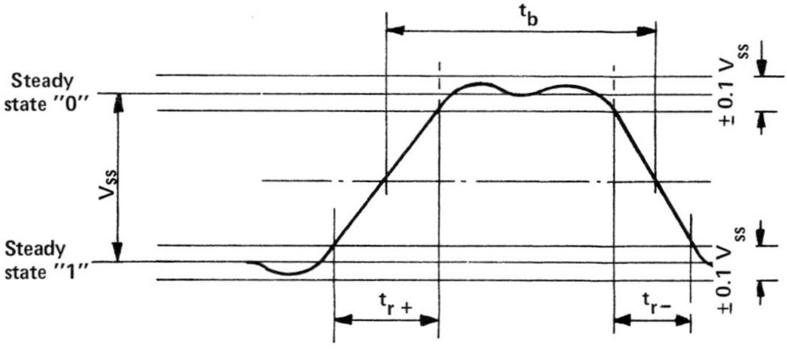

t_b = nominal duration of the test signal element

100 μs $\leqslant t_r \leqslant$ 300 μs when $t_b \geqslant$ 1 ms

0.1 $t_b \leqslant t_r \leqslant$ 0.3 t_b when $t_b <$ 1 ms

Figure 6.4 V.10 Generator Output Rise-time Measurement
(reproduced by kind permission of the ITU)

$V_{A'} - V_{B'} < -0.3v$	$V_{A'} - V_{B'} > +0.3v$
1 or OFF (Mark)	0 or ON (Space)

Table 6/3 V.10 Correlation

V.10 specifies several other electrical conditions to which the interface circuits must conform, which are not mentioned here.

Signal Waveform

The binary signal on the interchange circuits is deliberately rounded off to prevent interference with adjacent circuits (near-end crosstalk). Figure 6.4 is reproduced from V.10 and shows the restrictions on the pulse risetime.

Coaxial Cable

It is possible to use a coaxial cable for each interchange circuit instead of the normal multiconductor cable, but this requires generators with special characteristics.

Multipoint Operation

V.10 generators and receivers can be connected together in parallel (Figure 6.5). Receivers remain permanently 'on' and care has to be taken to ensure that only one generator is on at a time. The standard does not cover the protocol required to control data transfer.

Distance Limitations

A graph showing distance versus data rate for a typical V.10 implementation is shown in Figure 6.6.

This is reckoned to be a conservative guide, and 'in most practical cases the operating distances at the lower signalling rates may be extended to several kilometres'.

CCITT Rec V.11

V.11 operates up to 10 Mbit/s, using balanced interchange circuits. The equivalent circuit is shown in Figure 6.7. The voltage used is 6 volts maximum, and in fact the differential receiver is identical to the one used in V.10. The correspondence between electrical and logical conditions (Table 6/4) is also the same as for V.10.

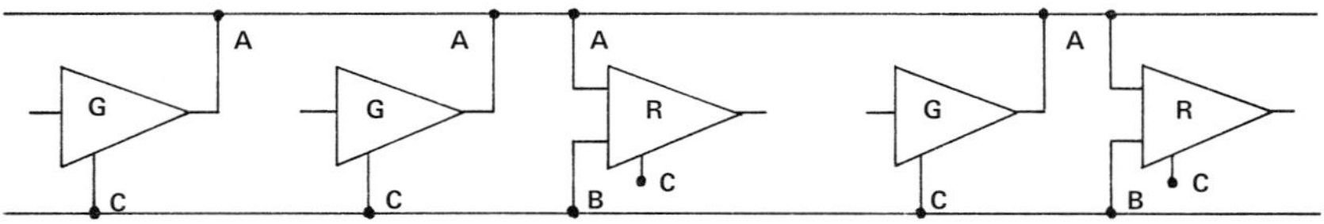

Figure 6.5 V.10 Multipoint Circuit

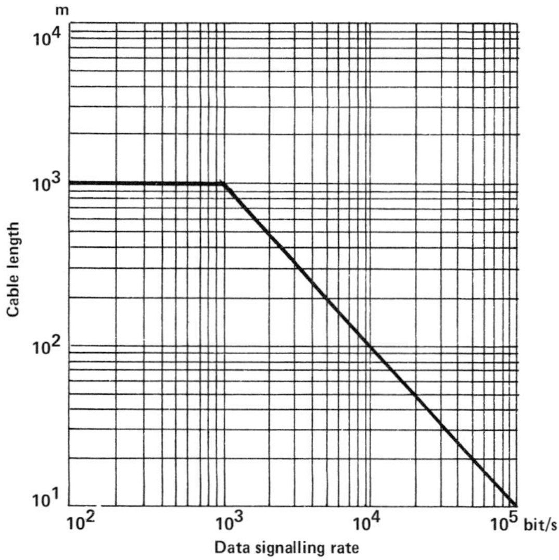

Figure 6.6 V.10 Data Signalling Rate vs Cable Length for Unbalanced Interchange Circuit

(Reproduced by kind permission of the ITU)

$V_{A'} - V_{B'} < -0.3v$	$V_{A'} - V_{B'} > +0.3v$
1 or OFF (Mark)	0 or ON (Space)

Table 6/4 V.11 Correlation

Signal Waveform

The binary signal is rounded off to reduce near-end crosstalk. Figure 6.8 is reproduced from V.11 and shows the restrictions on the pulse risetime.

Multipoint Operation

V.11 generators and receivers can be connected together in various multipoint configurations such as clusters, stars, or multidropped lines. As for V.10, precautions have to be taken to ensure that only one generator at a time is connected. No control protocols are specified for multipoint operation.

Distance Limitations

A graph showing distance versus data rate for a typical V.11 implementation is shown in Figure 6.9, for a terminated and unterminated cable. A cable is terminated by connecting a resistance of 100–150 ohms across it at the receiver. It is beneficial to do this at high data rates when the cable begins to act as a transmission line, since it helps to match the cable to the receiver and thus avoid reflections on the line.

As with V.10, Figure 6.9 is reckoned to be a conservative guide, and 'in many practical cases the operating distance at lower signalling rates may extend to several kilometres'.

CHOICE OF V.10 OR V.11

Only the higher speed circuits ever need to use V.11 electrical characteristics. These are the data, timing and certain control circuits, as listed below:

Circuit	103	Transmitted data	Data
	104	Received data	
	113	External transmit clock	Timing
	114	Transmit clock	
	115	Receive clock	
	105	Request to send	Control
	106	Clear to send	
	107	Data set ready	
	108	Data terminal ready	
	109	Carrier detect	

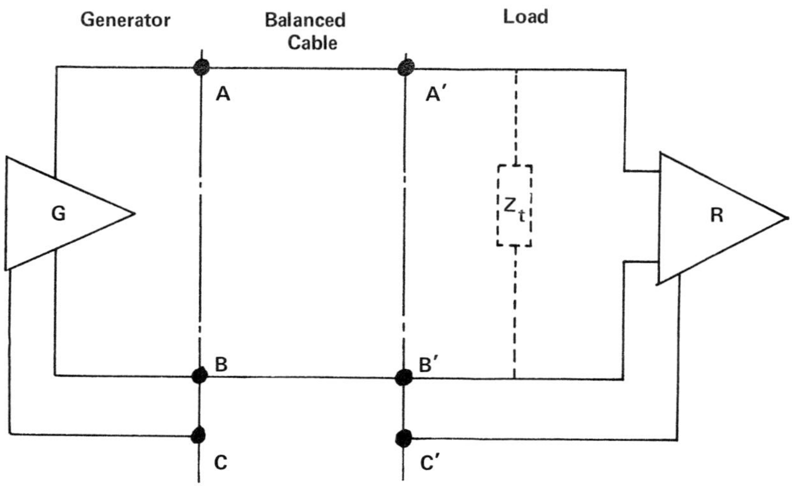

Figure 6.7 V.11 Equivalent Circuit

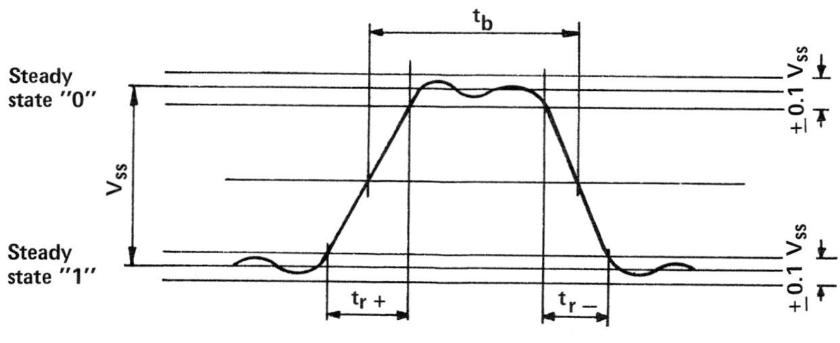

t_b = nominal duration of the test signal element

for $t_b \geqslant 200$ ns, $t_r \leqslant 0.1\ t_b$

for $t_b < 200$ ns, $t_r \leqslant 20$ ns

Figure 6.8 V.11 Generator Dynamic Balance and Rise-time Measurement (reproduced by kind permission of the ITU)

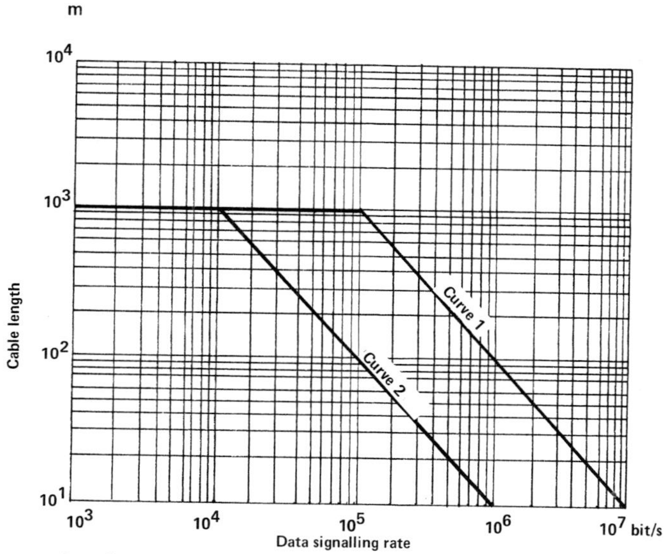

Curve 1: terminated interchange circuit

Curve 2: unterminated interchange circuit

Figure 6.9 V.11 Data Signalling Rate vs Cable Length for Balanced Interchange Circuit

(Reproduced by kind permission of the ITU)

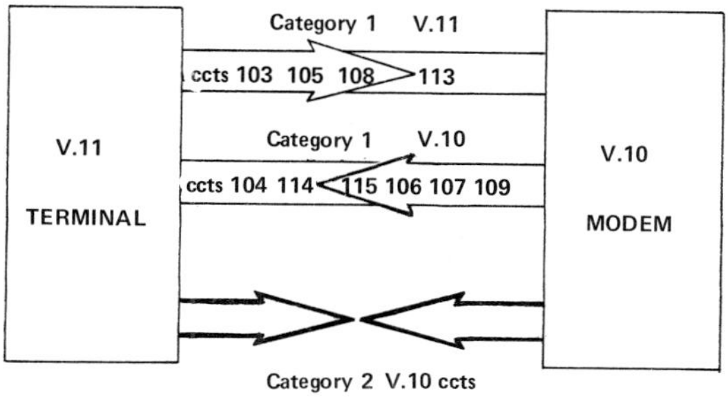

Figure 6.10 Interconnection of V.11 Terminal to V.10 Modem

The receivers for these circuits are known as category 1 receivers. In category 1 receivers, both receiver inputs (points A′ and B′ of Figure 6.7) are brought out to the interface connector. They can therefore be connected either to a balanced V.11 generator or, by grounding one of the inputs, to an unbalanced V.10 generator.

The normal criterion is that the above circuits use V.11 for operation above 20 kbit/s, and V.10 for operation up to 20 kbit/s. Exceptionally, V.11 may be used at low data rates to reduce interference or to permit longer cables to be used.

All circuits other than those in the above list have category 2 receivers, and they always conform to V.10. One of the receiver inputs is connected to a common return circuit in the equipment and is not available at the interface connector.

The above arrangement permits complete interworking between V.10 and V.11 equipment. Figure 6.10 shows an example, a V.11 terminal connected to a V.10 modem. The category 1 receivers in the V.10 modem are configured for V.11 balanced operation, and the category 1 receivers in the V.11 terminal configured for V.10 unbalanced operation. Signals on category 1 circuits therefore conform to V.11 in one direction, and V.10 in the other direction. Category 2 circuits all conform to V.10 anyway, and so there is no problem there.

7 Connectors

V.24 (V.28) AND RS232C INTERFACE

V.24 uses a 25-way D-shaped connector as shown in Figure 7.1. The socket (female contacts) is associated with the modem; the plug (male contacts) is associated with the terminal. Although the pin assignments have been standardised, the connector itself has still not become an official international standard, in spite (or perhaps because) of the existence of a manufacturer's *de facto* standard.

Pin Assignments – 25-way connector

The internationally agreed pin assignments are shown in Table 7/1, which is based on ISO 2110.* This International Standard leaves open certain assignments as national options, which leads to variations in the usage of some pins. Further complications can arise because RS232 predated ISO 2110 and is not wholly compatible with it. There is a number of pins which can cause problems, and Table 7/2 shows the different circuits which can be found on these pins.

CONNECTORS FOR AUTOMATIC CALLING

The same 25-way connector is specified in V.25 for the interchange circuits between the terminal and an automatic calling device, when used. The connector with female contacts (socket) is associated with the auto-calling equipment.

Pin assignments are well standardised; the US (RS366) and international (ISO 2110) standards are in agreement here. Table 7/3 lists the pin assignments and shows the nomenclature used for the 200-series interchange circuits.

* ISO 2110 : At the stage of a Draft International Standard (DIS) at the time of writing. (Revision of the 1972 issue.)

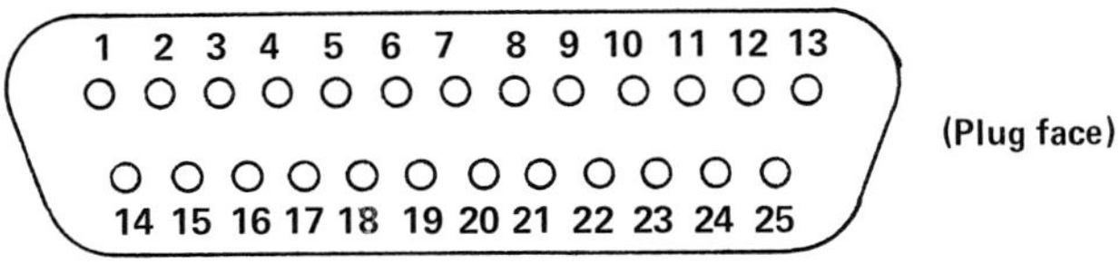

Figure 7.1 25-way Connector

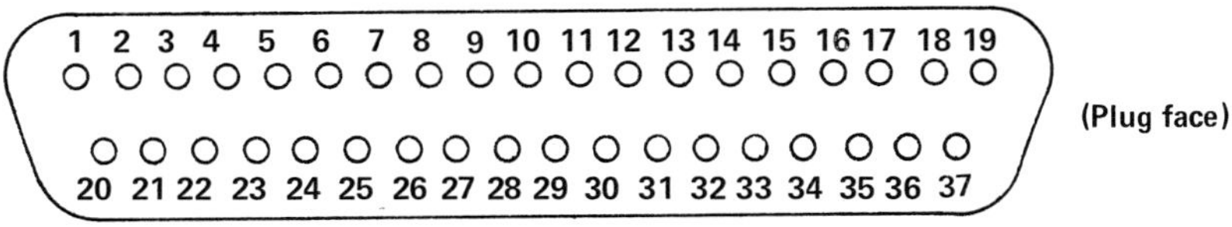

Figure 7.2 37-way Connector

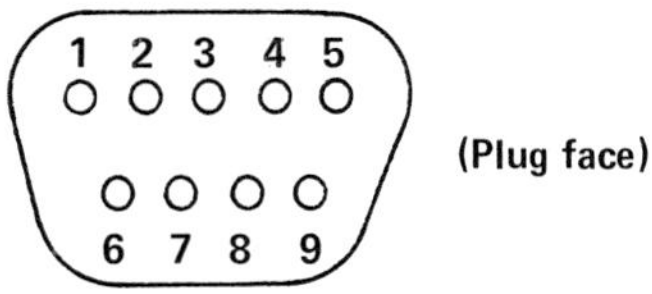

Figure 7.3 9-way Connector

V.24 (RS232) Circuit	Name	Pin Number
101 (AA)	Cable Screen	1
102 (AB)	Signal ground or common return	7
103 (BA)	Transmitted data	2
104 (BB)	Received data	3
105 (CA)	Request to send	4
106 (CB)	Ready for sending (Clear to send)	5
107 (CC)	Data set ready	6
108 (CD)	Connect data set to line/Data terminal ready	20
109 (CF)	Data channel received line signal detector	8
110 (CG)	Data signal quality detector	–
111 (CH)	Data signalling rate selector (DTE source)	23
113 (DA)	Transmitter signal element timing (DTE source)	24
114 (DB)	Transmitter signal element timing (DCE source)	15
115 (DD)	Receiver signal element timing (DCE source)	17
116	Select standby	(24)
118 (SBA)	Transmitted backward channel data	14
119 (SBB)	Received backward channel data	16
120 (SCA)	Transmit backward channel line signal	19
121 (SCB)	Backward channel ready	13
122 (SCF)	Backward channel received line signal detector	12
125 (CE)	Calling indicator	22
126	Select transmit frequency	11
140	Remote loopback for point-to-point circuits	21
141	Local loopback	18
142	Test indicator	25

Table 7/1 Pin Assignments in 25-Way Connector

Note: Modems for parallel data transmission use different pin assignments. See page 128.

PIN No	CIRCUITS			
	ISO DIS 2110 1979	UK PO	RS232C	Other
9	N	113	Test + 12V	
10	N	117	Test – 12V	
11	(126)	(111)	–	New synch/ remote loopback
15	114	113/114	114	
18	141	–	–	140
20	108.1 108.2	108.1 108.2	108.2	
21	140	–	110	
23	111	111	111/112	
24	113	116	113	
25	142	(117)	–	141/out of service

N – reserved for national use.

Table 7/2 Circuits Encountered on Certain Pins in the 25-Way Connector

V.24 (V.10/V.11) AND RS449

The use of V.11 balanced interchange circuits increases the number of wires in the interchange cable, since two leads per circuit are required. To accommodate the extra leads, two connectors are used at the interface – a 37-way connector for the main channel and associated circuits, and a 9-way connector for the backward channel circuits. For compatibility, the same connectors are used for V.10 unbalanced operation. These two connectors are from the same family as the 25-way connector, and have the same D-shape, (Figures 7.2 and 7.3).

37-Way and 9-Way Connectors

Pin assignments in these two connectors are agreed internationally, and given in ISO 4902.* The corresponding EIA Standard RS449 specifies the same pin assignments, although in RS449 new names have given to all the interchange

*ISO 4902 At the stage of a Draft International Standard (DIS) at the time of writing.

circuits. Nevertheless there should be less confusion over these two connectors than there has been over the 25-way connector.

Tables 7/4 and 7/5 give the pin assignments. (Appendix 3 gives a complete comparison of the nomenclature used in V.24, RS232C, and RS449).

Circuits CCITT V.24 (EIA RS366)	Designation	Pin No
201 (AB)	Signal ground or common return	7
202 (CRQ)	Call request	4
203 (DLO)	Data line occupied	22
204 (COS)	Distant station connected (call origination status)	13
205 (ACR)	Abandon call	3
206 (NB1)	Digit signal 2^0	14
207 (NB2)	Digit signal 2^1	15
208 (NB4)	Digit signal 2^2	16
209 (NB8)	Digit signal 2^3	17
210 (PND)	Present next digit	5
211 (DPR)	Digit present	2
213 (PWI)	Power indication	6

Table 7/3 Pin Assignments in 25-Way Connector Used For Automatic Calling

V.24 Circuit	CCITT Designation	Pin
102	Signal ground or common return	19
102a	DTE common return	37
102b	DCE common return	20
103	Transmitted data	4 + 22
104	Received data	6 + 24
105	Request to send	7 + 25
106	Ready for sending (CTS)	9 + 27
107	Data set ready	11 + 29
108	Connect data set to line/ Data terminal ready	12 + 30
109	Data channel received line signal detector	13 + 31
110	Data signal quality detector	33*
111	Data signalling rate selector (DTE source)	16
112	Data signalling rate selector (DCE)	2*
113	Transmitter signal element timing (DTE source)	17 + 35
114	Transmitter signal element timing (DCE source)	5 + 23
115	Receiver signal element timing (DCE source)	8 + 26
116	Select standby	32*
117	Standby indicator	36*
125	Calling indicator	15
126	Select transmit frequency	16
135	Terminal available for service	28*
136	New signal	34*
140	Remote loopback for point to point circuits	14
141	Local loopback	10
142	Test indicator	18

*These are the preferred assignments on pins available for national use.

Table 7/4 V.10/V.11 37-Way Connector – Pin Assignments (Based on ISO DIS 4902)

V.24 Circuit	CCITT Designation	Pin No
102	Common return/signal ground	5
102a	DTE common return	9
102b	DCE common return	6
118	Transmitted backward channel data	3
119	Received backward channel data	4
120	Transmit backward channel line signal	7
121	Backward channel ready	8
122	Backward channel received line signal detector	2
–	Cable screen	1

Table 7/5 V.10/V.11 9-Way Connector – Pin Assignments (Based on ISO DIS 4902)

INTERWORKING BETWEEN OLD (V.28/RS232) AND NEW (V.10/RS449) STANDARDS

The problems of interworking fall into two classes – electrical and mechanical.

Electrical: There is an overlap in the electrical characteristics of V.28 and V.10 in the region between 5 and 6 volts and, in practice, the V.28 equipment will probably respond satisfactorily to a voltage of only 4 volts. This permits interworking between the old and new equipment. Points to note are:

- the interchange circuit length and the maximum data rate are constrained to the V.28 limits (15 metres, 20 kbit/s);
- the new equipment must use V.10 interchange circuits;
- the new equipment may only be designed to withstand voltages up to 12v, whereas V.28 equipment can produce 25 volts. A pair of resistors across each interchange circuit provides the necessary protection for the V.10 equipment. To keep down near-end crosstalk, they need to be placed within 3 metres of the V.10 receivers. Wiring diagrams are given in Appendix 2.

Mechanical: The incompatibilities between connectors can be overcome by using a simple adaptor cable or adaptor box.

V.35 INTERFACE

The connector specified for the V.35 48 kbit/s balanced interface is a 34-way square connector, as sketched in Figure 7.4. The pin assignments shown in Table 7/6 are based on international standard ISO 2593: 1973; there is no corresponding EIA standard.

CCITT Circuit	Name	Pin ISO*
101		A
102	Common return or signal ground	B
103	Transmitted data	P & S
104	Received data	R & T
105	Request to send	C
106	Ready for sending (clear to send)	D
107	Data set ready	E
108	Connect data set to line/Data terminal ready	H
109	Carrier detect	F
113	Transmit timing (external clock)	U & W
114	Transmit timing	Y & AA(a)
115	Receive timing	V & X
125	Calling indicator	J
—	Transmitter clock control	HH(g)

* The UK pin labelling is shown in brackets, where it differs from ISO labels.

Table 7/6 34-Way Connector Pin Allocations

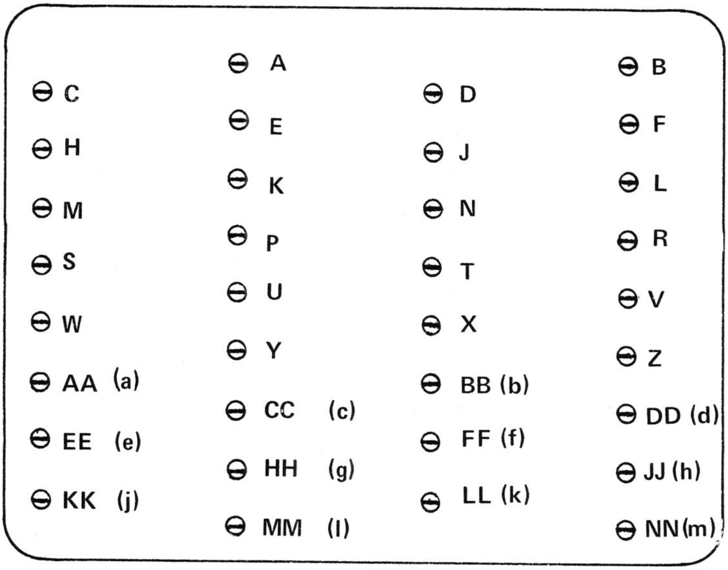

Figure 7.4 34-pin Connector Layout
Letter in brackets show where UK pin labelling differs from ISO labels

8 Modem Variants and Ancillary Equipment

BASEBAND MODEMS

A baseband modem does not modulate or demodulate a carrier signal in the way that a true modem does. It merely takes the binary data signal and transmits it to line in a modified form, as a baseband signal. (A baseband signal, it will be recalled, is a signal containing frequencies down to dc.) The modification takes the form of pulse shaping – rounding off the square pulses to reduce the high-frequency components – and scrambling, to give the baseband signal a satisfactory frequency spectrum independent of the binary data signal.

Baseband modems require a physical pair of wires over which to transmit, since the baseband signal contains frequencies outside the normal 300–3400 Hz speech band. They are therefore only suitable for short-haul applications where an unloaded physical pair is available.

One particular application of baseband transmission is found in wideband 48 kbit/s links, for which special cable capable of transmitting the high-frequency signals has to be laid to the customer's premises. Short point-to-point links can be cabled directly, and baseband modems would be used at each end of the link. Longer point-to-point circuits are routed via 48 kbit/s circuits derived from the telephone FDM system. In this situation, baseband transmission is used for the two ends of the circuit, between the customer and the group band modem located at the nearest trunk network access point (Figure 8.1).

ACOUSTIC COUPLERS

Basically, an acoustic coupler is a modem equipped with audio transducers (a loudspeaker and a microphone) so that it can interface to the telephone handset instead of to the telephone line. It can be used at low data rates where the standard modulation method is frequency shift keying.

An acoustic coupler presents a normal V.24 interface to the terminal. It

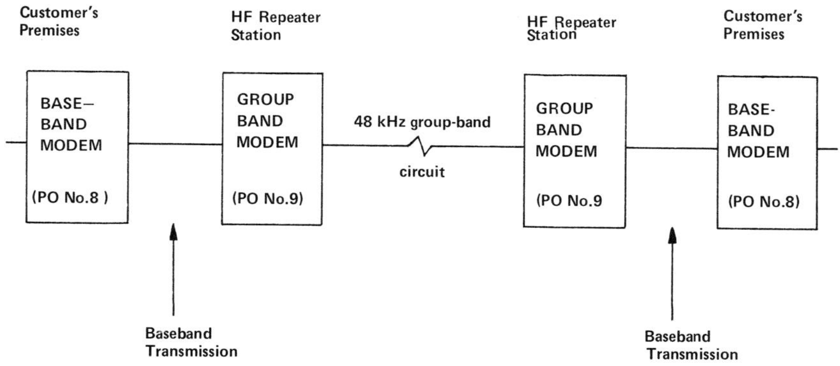

Figure 8.1 Modems in a 48 kbit/s Link

accepts data from the terminal and converts it into audible high- and low-frequency tones, which are then fed to the microphone in the telephone hand-set. In the reverse direction, the coupler converts the audible tones from the telephone earpiece into binary data signals for transfer to the terminal. Some terminals have built-in acoustic couplers.

The great advantage of an acoustic coupler is that it permits a terminal to use any convenient telephone. No wiring or other modifications are necessary. Its performance is inferior to a normal modem – the error rate is worse – but this is not a serious problem for many users.

Acoustic couplers are (reluctantly) permitted by the PTTs, and have to be approved to ensure they do not interfere with other users or with telephony plant. They are not favoured for permanent terminal installations.

LIMITED-DISTANCE MODEMS

The modems standardised by the CCITT are designed for international and intercontinental use. Over shorter distances it is possible to achieve equivalent performance using less sophisticated modulation/demodulation and equalisation techniques, and there is now a number of limited-distance modems offering very good price/performance ratios. Typically these modems have a range of about 50 miles, and operate up to 4800 bit/s.

Note, however, that not all 'short-haul' or 'limited-distance' modems are true modems. Many are baseband modems and can only be used when unloaded physical pairs are available, which limits their range considerably.

MODEM ELIMINATORS

Modem eliminators are employed for in-house transmission up to a mile or so, where the use of full modems is not justified. A modem eliminator is a line driver/receiver packaged to look like a modem and providing the same V.24 circuits as a modem. The line driver amplifies or regenerates the interface signals for transmission over twisted pairs or coaxial cable.

DOUBLE DIAL-UP AND SPLIT-STREAM MODEMS

Double dial-up modems are designed for use on the public switched telephone network. Two telephone connections are established to the required destination, and the two paths are then used as a 4-wire connection providing full-duplex transmission (Figure 8.2).

Split-stream modems also use two separate circuits, but in this case both circuits carry data in the same direction. By transmitting at 9.6 kbit/s over two high grade leased lines operating in parallel, a split-stream modem can provide a data throughput rate of 19.2 kbit/s (Figure 8.3).

The electronic circuitry which splits the data stream and handles all the synchronisation problems can take the form of a separate unit, when it is known as a lineplexer.

MODEM-SHARING UNIT

A modem-sharing unit, also called a fan-out unit, allows several terminals to share one modem. All the terminals receive data from the modem simultaneously, but only one terminal can transmit at any one time. This distinguishes the device from a multiplexing modem which provides individual channels to each terminal it serves.

Modem-sharing units are economical where several colocated terminals require access to the same information and where the data traffic generated by each is low. They may also be used at a computer centre to permit more than one computer port to have access to a communications link, which can be useful in circumventing faults.

ANALOGUE MODEMS

Instrumentation systems generate analogue signals which can vary continuously within a given range of values.

Transmission of analogue data requires a modem that can convey such a continuously varying signal, in contrast to the simple two-state signal found

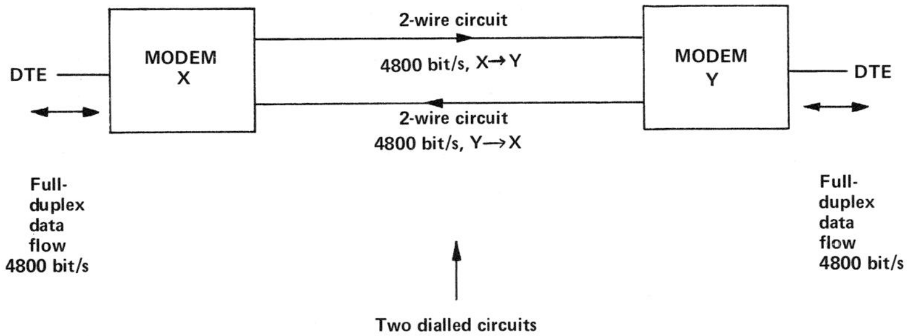

Figure 8.2 Double Dial-up (4800 bit/s)

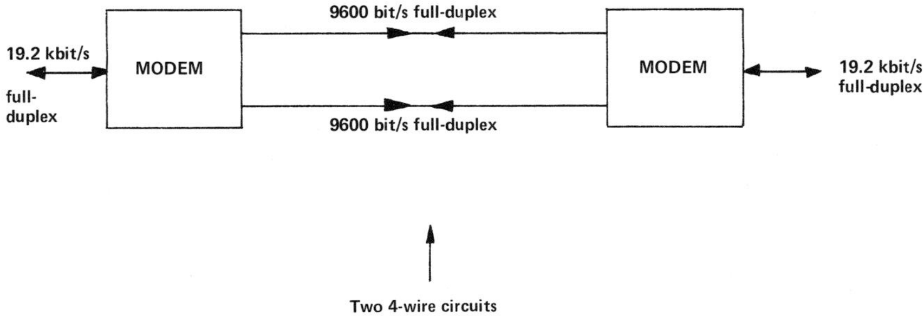

Figure 8.3 Split-stream Modem (19.2 kbit/s)

in digital data transmission.

CCITT has standardised one analogue modem for use in medical applications (Recommendation V.16). It uses frequency modulation, the frequency of the transmitted carrier varying linearly with the voltage applied to the modem. Voltages between −2.5 and +2.5 volts are used. The bandwidth of the analogue signal can be up to 100 Hz. The modem can also be used for digital data, when the modulation system becomes straightforward frequency shift keying. Three independent channels are provided centred on the frequencies 950, 1400 and 2100 Hz, together with a 200 baud digital backward channel.

MODEMS FOR PARALLEL DATA TRANSMISSION

Parallel data transmission employs simultaneous tones to transmit one character at a time.

The simplest type of parallel 'modem' is the multi-frequency push-button telephone, which transmits a different pair of frequencies for each button pressed. There are 16 different frequency pairs, and hence 16 different characters. Combining characters in different sequences increases the range of meanings that can be conveyed and gives an adequate repertoire for applications such as order entry, where the push-button telephone serves as a very economical outstation.

The receiving modem at the central instation uses a special type of V.24 interface which has 8 Received Data interchange circuits corresponding to the 8 frequencies used. When the modem detects a pair of frequencies on the line, it sets the appropriate Received Data circuits to binary 1. The instation may transmit a 420 Hz tone or a voice response; provision is made for both at the interface.

A system of this type has been standardised and is specified in CCITT Rec V.19. A more sophisticated international system has also been specified by CCITT, in Rec V.20. This uses a modem at the outstation instead of a push-button telephone. There are 16 basic character combinations, but provision is made for increasing this to 64 by using 4 extra tones. The maximum character transfer rate is 40 characters per second. Provision is made for a simultaneous backward channel, which can be a 420 Hz tone for audible signalling or a V.23 75 bit/s backward data channel. Alternatively voice response can be employed.

The interchange circuits used are listed in Table 8/1. The instation uses V.28 electrical characteristics. The outstation uses an ON/OFF current loop, defined in CCITT Rec V.31; the data circuits have a common return circuit, but the control circuits all have separate return circuits.

V.24 Circuit	Designation	25-way connector Pin assignments	
		Instation	Outstation
102	Signal ground or common return	24	13
103	Transmitted data (9 or 6 circuits)	–	3-5, (6-8), 9-11
104	Received data (12 or 8 circuits)	3-6, 9-12, (13-16)	–
105	Request to Send	20	16 & 17
107	Data set ready	23	(22 & 23)
108	Connect data set to line/Data terminal ready	22	(24 & 25)
109	Data channel received line signal detector	8	–
110	Data signal quality detector	(2)	–
119	Received backward channel data		(20 & 21)
124	Select frequency groups	(25)	–
125	Calling indicator	21	(14 & 15)
129	Request to receive		18 & 19
130	Transmit backward tone	19	–
131	Received character timing	(7)	–
191	Transmitted voice answer	17 & 18	–
192	Received voice answer	–	(2 & 12)
	Cable screen	1	1

Pin numbers in brackets indicate that provision of this circuit is optional.

Table 8/1 V.20 Interchange Circuits, For Parallel Data Transmission

VOICE ADAPTER

A voice adaptor is an option available for some modems which permits a private leased circuit to be used for speech when it is not being used for data. The adapter takes the form of a telephone handset connected to the modem, together with the necessary switching facilities to connect the telephone to line, and signalling facilities for calling the distant telephone.

The Post Office can provide equivalent facilities where the modem supplier does not offer a voice adapter option. (On PSTN connections alternative voice facilities are always provided by the Post Office).

MULTIPLEXERS AND CONCENTRATORS

Multiplexers and concentrators allow a number of data channels to be carried on a single high-speed data link. Both devices offer savings in line costs by permitting one circuit to be used instead of several.

Traditionally concentrators employed contention whereas multiplexers did not, but this distinction is becoming blurred as multiplexers become more intelligent. Concentrators are programmable (usually they are minicomputers), and they can therefore reduce the load on the central mainframe in a computer network by performing tasks such as polling, protocol conversion or system reconfiguration remotely.

TEST EQUIPMENT

Interface Tester

An interface tester is a small portable unit used to monitor the conditions at the V.24 interface. It is connected between the terminal and the modem, and indicates the states of all the interchange circuits by means of light-emitting diodes. Facilities are provided for strapping circuits together, or to a permanent ON or OFF condition.

Test Set

A test set measures bit error rate and is used for testing and commissioning data transmission circuits. Typically the circuit to be tested is looped back at the remote modem and the tester connected to the local modem. The tester transmits a 511- or 2047-bit pseudo-random test pattern at the selected data rate, and measures the bit error rate and distortion in the returned bit sequence.

Level Measuring Set

A level measuring set measures the power of an analogue signal. It allows the attenuation (insertion loss) of a circuit to be checked, and is especially useful for setting up multipoint circuits. It also permits the level of the input and output signals on the line side of the modem to be measured.

Data Link Analyser

Data link analysers are a very powerful (but expensive) aid to fault diagnosis. They are sophisticated microprocessor-controlled testers which permit on-line monitoring of data link protocol. Many make provision for digital recorders so that observed sequences can be stored for subsequent play-back.

The analyser has a screen for displaying the significant events on the link, and a keyboard via which various tests can be programmed by the operator. Units can accept a variety of codes and protocols, and may be able to emulate terminals or CPUs in a polled environment.

Other Test Equipment

Psophometers (for measuring white noise), impulse counters, group delay measuring sets and frequency counters are among the other types of test equipment available, but none is likely to prove as useful as the items outlined above. The only other piece of test equipment that should not be forgotten is a pair of headphones.

9 Data Networks

The design of data networks is not a topic that can be tackled in the closing chapter of this book. Factors involved in the choice of leased lines vs the PSTN, modem-sharing units vs multiplexers, multipoint vs point-to-point circuits etc are too many and complex to cover here.

Suffice it to note that the emerging public switched data networks could relieve the user of many difficult decisions in this area. Instead of working out the optimum topology for a private network, planning the locations of multiplexers and concentrators, the computer network planner merely has to arrange for all terminals and computer mainframes to be connected to the public network. Expansion or reconfiguration of the user network to serve changing needs are no problem. Likewise management and maintenance of the network are the responsibility of the PTT.

It will be interesting to see over the next decade whether PTTs can make public data networks sufficiently attractive in terms of reliability, security and cost, to persuade users to migrate from their private networks to the new public networks.

INTERFACES FOR PUBLIC DATA NETWORKS

The services to be provided on the new public data networks include digital leased lines, circuit switched services and packet switched services. At the lowest level – the level of the DTE-DCE interchange circuits – the interface is the same for all services, and is somewhat simpler than a V.24 interface. This is because many of the functions are performed at the logical rather than the physical level.

In a digital network, the DCE is not a modem as it would be in an analogue network, but a form of coder/decoder, which takes the binary data from the terminal and converts it to a form suitable for digital transmission along the line. It is an unintelligent device. The intelligence resides at the data switching exchange, and to set up a call on a public data network, a terminal must communicate with the exchange. The terminal-exchange protocol is one component of the DTE-DCE interface specification.

There are two basic interface specifications, defined in CCITT recommendations X.20 and X.21. X.20 is designed for start/stop terminals and provides full-duplex transmission up to 300 bit/s. X.21 is designed for synchronous terminals, and provides full-duplex operation up to 48 kbit/s. (X.25, the interface to a packet switched network, is a three-level user-to-packet network interface protocol; it uses X.21 at its lowest level.)

With these new interfaces, the terminal is directly responsible for all aspects of call establishment and disconnection. A handshake protocol is defined in which the terminal inputs the number (address) of the party to be called, together with details of any special facilities, and the data exchange responds with instructions and call progress signals. Call establishment time is less than one second.

X.20 DCEs employ the electrical characteristics of Rec V.10 (= X.26); they may be connected to DTEs using V.10, V.11 or V.28 electrical characteristics. X.21 DCEs employ V.11 (= X.27); below 9600 bit/s the DTE may use V.10 or V.11 electrical characteristics, but above 9600 bit/s only V.11 DTEs (with cable termination in the load) are permitted. Both use a selection of interchange circuits from Rec X.24 (the equivalent of V.24), and a 15-way D-shape connector. Table 9/1 shows the interchange circuits used and connector pin assignments for the two standards based on ISO DIS 4903 (1979).

Circuit	Name	Direction	X.20	X.21	15-Way Connector Pin Number
G	Signal ground or common return	–	X	X	8
Ga	DTE common return	to DCE	(X)	(X)	9 (+10)
Gb	DCE common return	to DTE	(X)		11
T	Transmit	to DCE	X	X	2 (+9)
R	Receive	to DTE	X	X	4 (+11)
C	Control	to DCE		X	3 +10
I	Indication	to DTE		X	5 +12
S	Signal element timing	to DTE		X	6 +13
B	Byte timing	to DTE		(X)	7 +14
	Cable screen				1

(Brackets indicate optional features)

Table 9/1 – X.20 and X.21 Interchange Circuits

Recognising the need for the new networks to cater for the existing population of V.24 terminals, CCITT has developed alternative versions of the two basic standards; they are known as X.20 *bis* and X.21 *bis*. They employ V.24 interchange circuits with the traditional electrical characteristics and connectors, and thus make the digital network DCE look like a modem to the terminal. They do not offer all the facilities of the full X.20 and X.21 interfaces.

Appendix 1

Availability of Standards

CCITT RECOMMENDATIONS

CCITT recommendations are designated by a letter and a number. All the recommendations dealing with data transmission over the telephone network have numbers in the V-series. Recommendations are published in the CCITT 'Book', which is published every 4 years under a different colour cover each time. The 1976 CCITT book is the Orange Book, and the V-series recommendations are contained in Volume VIII.1. The titles are listed below.

V.1 Equivalence between binary notation symbols and the significant conditions of a two-condition code.

V.2 Power levels for data transmission over telephone lines.

V.3 International Alphabet No. 5.

V.4 General structure of signals of International Alphabet No. 5 code for data transmission over public telephone networks.

V.5 Standardization of data-signalling rates for synchronous data transmission in the general switched telephone network.

V.6 Standardization of data-signalling rates for synchronous data transmission on leased telephone-type circuits.

V.10 Electrical characteristics for unbalanced double-current interchange circuits for general use with integrated circuit equipment in the field of data communications.

V.11 Electrical characteristics for balanced double-current interchange circuits for general use with integrated circuit equipment in the field of data communications.

V.15 Use of acoustic coupling for data transmission.

V.16 Medical analogue data transmission modems.

V.19 Modems for parallel data transmission using telephone signalling frequencies.

V.20 Parallel data transmission modems standardized for universal use in the general switched telephone network.

V.21 200-baud modem standardized for use in the general switched telephone network.

V.23 600/1200-baud modem standardized for use in the general switched telephone network.

V.24 List of definitions for interchange circuits between data-terminal equipment and data circuit-terminating equipment.

V.25 Automatic calling and/or answering equipment on the general switched telephone network, including disabling of echo-suppressors on manually established calls.

V.26 2400 bits per second modem standardized for use on 4-wire leased telephone-type circuits.

V.26 *bis* 2400/1200 bits per second modem standardized for use in the general switched telephone network.

V.27 4800 bits per second modems with manual equalizer standardized for use on leased telephone-type circuits.

V.27 *bis* 4800 bits per second modem with automatic equalizer standardized for use on leased telephone-type circuits.

V.27 *ter* 4800/2400 bits per second modem standardized for use in the general switched telephone network.

V.28 Electrical characteristics for unbalanced double-current interchange circuits.

V.29 9600 bits per second modem standardized for use on leased telephone-type circuits.

V.31 Electrical characteristics for single-current interchange circuits controlled by contact closure.

V.35 Data transmission at 48 kilobits per second using 60-108 kHz group band circuits.

V.36 Modems for synchronous data transmission using 60-108 kHz group band circuits.

V.40 Error indication with electromechanical equipment.

V.41 Code-independent error control system.

V.50 Standard limits for transmission quality of data transmission.

V.51 Organization of the maintenance of international telephone-type circuits used for data transmission.

V.52 Characteristics of distortion and error-rate measuring apparatus for data transmission.

V.53 Limits for the maintenance of telephone-type circuits used for data transmission.

V.54 Loop test devices for modems.

V.55 Specification for an impulsive noise measuring instrument for telephone-type circuits.

V.56 Comparative tests of modems for use over telephone-type circuits.

V.57 Comprehensive data test set for high data signalling rates.

Volume V111.2 contains the X-series recommendations and the list below gives the titles of relevant recommendations.

X.20 Interface between data terminal equipment (DTE) and data circuit-terminating equipment (DCE) for start-stop transmission services on public data networks.

X.20 *bis* V.21-compatible interface between data terminal equipment (DTE) and data circuit-terminating equipment (DCE) for start-stop transmission services on public data networks.

X.21 General purpose interface between data terminal equipment (DTE) and data circuit-terminating equipment (DCE) for synchronous operation on public data networks.

X.21 *bis* Use on public data networks of data terminal equipments (DTEs) which are designed for interfacing to synchronous Series V Recommendations modems.

X.24 List of definitions for interchange circuits between data terminal equipment (DTE) and data circuit-terminating equipment (DCE) on public data networks.

Volumes of the Orange Book are available from ITU sales, Place des Nations, CH-1211 Geneva 20, Switzerland, (telephone Geneva 9951 11).

In the UK, a complete set of CCITT Books is available for reference at the Library of the Institution of Electrical Engineers, Savoy Place, London WC2R 0BL (telephone 01-240 1871).

The 1980 CCITT book is due for publication in 1981.

EIA RECOMMENDED STANDARDS

The Electronic Industries Association is an American trade association representing manufacturers of electronic products. It is active in producing standards for the electronics industry, and many EIA standards are subsequently adopted as American National Standards. EIA recommended standards on data communications are listed below.

Data Rate

RS-269-B (1976) Synchronous signalling rates for data transmission.

DTE-DCE Interface

RS-232-C (1969) Interface between DTE and DCE employing serial binary data interchange.

EIA Industrial Electronics Bulletin No. 9 'Application notes for EIA standard RS-232-C' (1971) is a recommended companion document.

RS-334 (1967) Signal quality at interface between DTE and synchronous DCE for serial data transmission.

RS-363 (1969) Standard for specifying signal quality for transmitting and receiving DTEs using serial data transmission at the interface with non-synchronous DCE.

EIA Industrial Electronics Bulletin No. 5 'Tutorial Paper on signal quality at digital interface' 1965.

RS-366 (1969) Interface between DTE and automatic calling equipment for data communication.

RS-404 (1973) Standard for start-stop signal quality between DTE and non-synchronous DCE.

RS-410 (1974) Standard for the electrical characteristics of class A closure interchange circuits.

RS-422-A (1978) Electrical characteristics of balanced voltage digital interface circuits.

RS-423-A (1978) Electrical characteristics of unbalanced voltage digital interface circuits.

RS-449 (1977) General purpose 37-position and 9-position interface for DTE and DCE employing serial binary data interchange.

EIA Industrial Electronics Bulletin No. 12 'Application notes on interconnection between interface circuits using RS-449 and RS-232-C' (1977) is a useful companion document.

Fault Isolation

EIA Industrial Electronics Bulletin No. 11 'Fault isolation methods for Data Communication Systems' 1972.

Standards and a free catalogue are available from Electronic Industries Association, Engineering Department – Standards Sales Office, 2001 Eye Street NW, Washington DC 20006, USA. Tel +1 202 457-4963.

INTERNATIONAL STANDARDS ORGANISATION (ISO)

ISO standards are available via the national standards body, which in the UK is the British Standards Institution. The ISO standards for connectors are listed below.

ISO 2110 : 1979 Data Communications. DTE-DCE Interchange circuits; Assignment of Connector Pin Numbers.

ISO 2593 : 1973 Connector Pin allocations for use with high speed data terminal equipment.

ISO 4902 : 1979 37-pin and 9-pin DTE-DCE Interface Connectors and pin assignments.

ISO 4903 : 1979 Data communications. 15-pin DTE-DCE Interface Connector and pin assignments.

UK POST OFFICE TECHNICAL GUIDES

Technical requirements for equipment connected to Post Office plant or equipment are listed in a series of technical guides published by the Post Office, and are shown below.

No. 1 (Dated September 1973)	Post Office Datel Services – Datel 100 Service. The connection of customers' equipment to Post Office Telegraph (including Telex) lines, circuits and equipment in the United Kingdom.
No. 2 (Dated March 1971 – Reissued October 1976)	Post Office Datel Services – General requirements for data terminal equipment.
No. 3 (Dated October 1976)	Datel 200 Service. Connection of data terminal equipment to Post Office Datel Modem No. 2B.
No. 5 (Dated September 1975 – Reissued October 1976)	Datel 600 Service. Connection of data terminal equipment to Post Office Datel Modems.
No.6 (Dated March 1976)	Datel 2400 Service. Connection of data terminal equipment to Post Office Datel Modem No. 7.
No.8 (Dated November 1976)	Automatic calling equipment. The connection of data terminal equipment to the Data Control Equipment No. 1A.

No. 9 (Dated March 1971 – Reprinted March 1977)	Automatic answering. Connection of data terminal equipment to the Post Office Data Control Equipment No. 2A.
No. 10 (Dated November 1974 – Reissued December 1976)	Post Office Datel Services. Data Control Equipment No. 3A. Automatic origination and answering of calls on Telex.
No. 13 (Dated March 1977)	Datel 2412 Service. Connection of data terminal equipment to Post Office Modem 12.
No. 14 (Dated September 1975 – Reprinted March 1977)	Datel 200 Service. Connection of Data Terminal Equipment to the PO Datel Modem No. 13A.
No. 15 (Dated September 1979)	Datel 4800 Service. Connection of Data Terminal Equipment to the Post Office Modem 11.
No. 18 (Dated July 1978)	Datel 4832 Service. Connection of Data Terminal Equipment to PO Datel Modems.
No. 25 (Dated March 1976 – Reprinted December 1976)	Transmission design requirements for the connection of privately owned and maintained equipments to Post Office private circuits.
No. 26 (Dated March 1976 – Reissued March 1977)	Protection design requirements for the connection of privately maintained equipment or wires to Post Office maintained equipment or circuits.
No. 30 (Dated March 1976 – Reissued March 1977)	Requirements for attachments connected to the public switched telephone network.
No. 32 (Dated July 1975 – Reprinted October 1976)	Acoustic Coupling – General requirements and use of acoustically coupled modems for data transmission over Post Office lines.
No. 35 (Dated October 1975 – Being revised)	Requirements for attachments connected to Post Office and privately provided PBXs.

Technical Guides are available from:

Post Office Telecommunications Headquarters ME/RCS 1.2.1.4,
Tenter House, 45 Moorfields, LONDON EC2Y 9TH.
(01-432 9347)

Appendix 2 Connector Wiring Diagrams

1 DTE-DTE Asynchronous Terminal to Computer, 4-wire

2 DCE-DCE Modem to Modem, Back to Back

3 DTE-DTE Synchronous Terminal to Computer

4 DTE-DTE Synchronous Terminal to Synchronous Terminal

5 Modem Loopback (digital)

6 Terminal Loopback

7 Loopback Timing

8 Secondary Forward Channel

9 Adaptor, V.28 Modem to V.10 Terminal

10 Adaptor, V.28 Terminal to V.10 Modem

Pin	Circuit	CCITT	(EIA)
1	Cable screen	101	(AA)
2	Transmit data	103	(BA)
3	Received data	104	(BB)
4	Request to send	105	(CA)
5	Ready for sending/Clear to send	106	(CB)
6	Data set ready	107	(CC)
7	Common return	102	(AB)
8	Carrier detect	109	(CF)
17	Receive clock	115	(DD)
20	Data terminal ready	108/2	(CD)
24	External transmit clock (to modem)	113	(DA)

Note: External transmit clock (circuit 113) may be found on pins other than pin 24. See Table 7/2 page 116.

ASYNCHRONOUS TERMINAL to COMPUTER (DTE-DTE)

[extended over 4-wire circuit.]

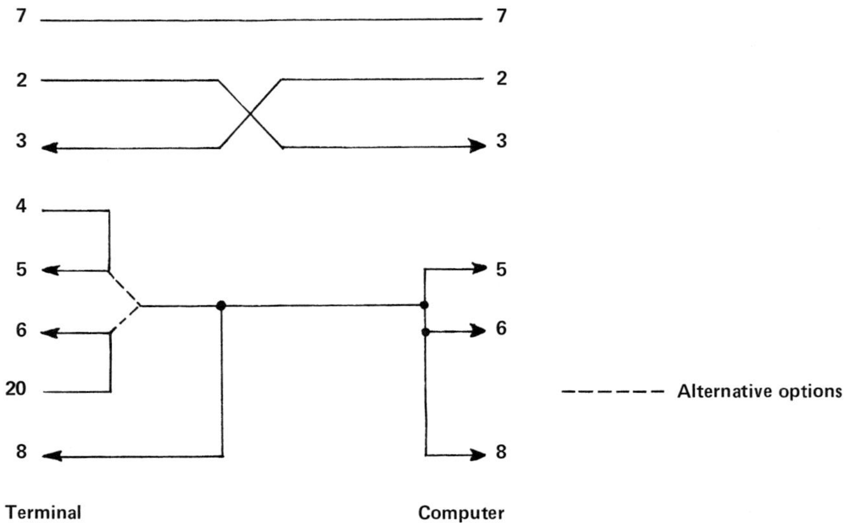

Appendix 2

MODEM to MODEM (DCE–DCE)

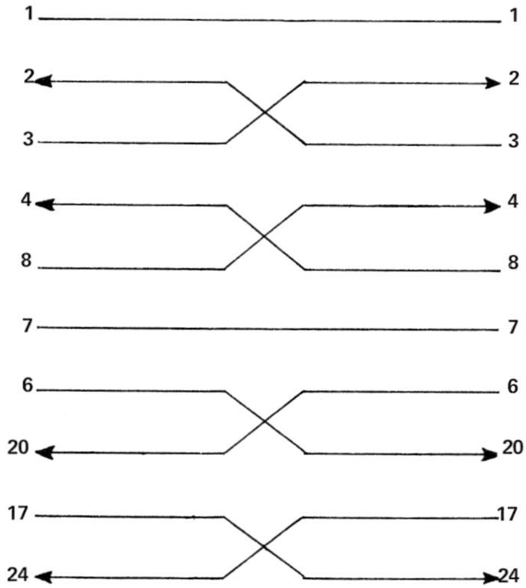

Modems must be strapped for external transmit clock

SYNCHRONOUS TERMINAL to COMPUTER direct
(clock in computer)

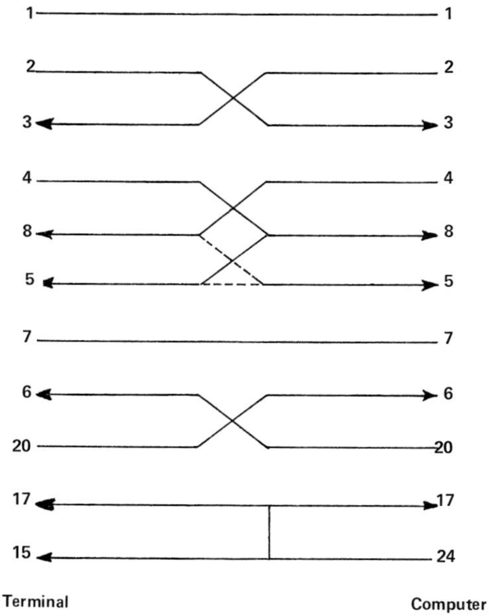

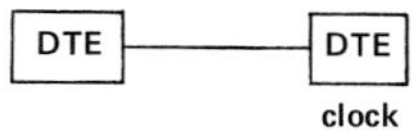

– – – –Alternative options

Pin	Circuit	CCITT	(EIA)
2	Transmit data	103	(BA)
3	Received data	104	(BB)
4	Request to send	105	(CA)
5	Ready for sending/Clear to send	106	(CB)
6	Data set ready	107	(CC)
8	Carrier detect	109	(CF)
15	Transmit clock (from modem)	114	(DB)
17	Receive clock	115	(DD)
20	Data terminal ready	108/2	(CD)
24	External transmit clock (to modem)	113	(DA)

Note: External Transmit clock (circuit 113) may be found on pins other than pin 24. See Table 7/2, page 116.

SYNCHRONOUS TERMINAL TO SYNCHRONOUS TERMINAL
(clock supplied from external source)

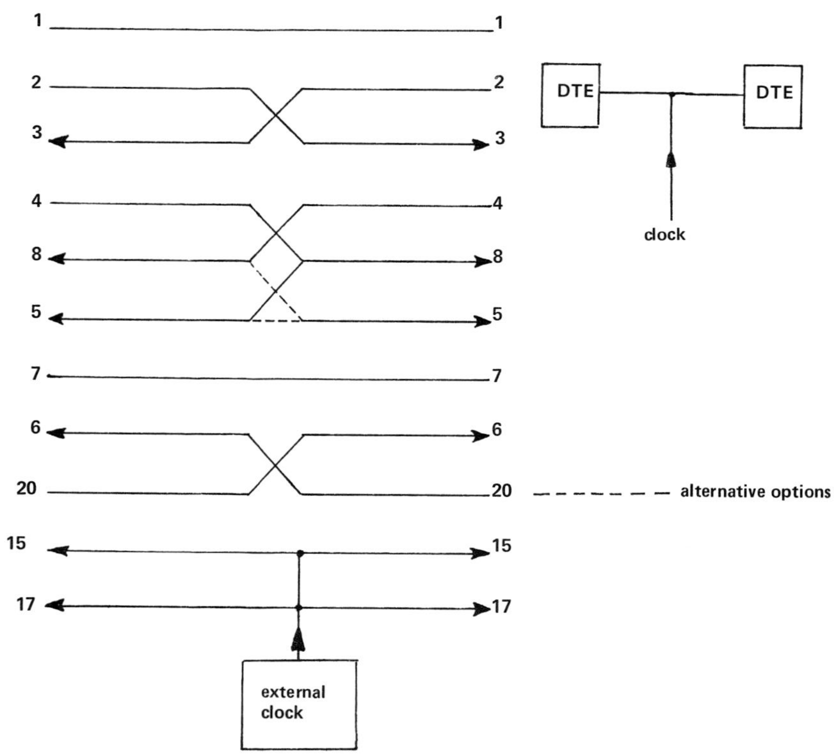

Appendix 2

MODEM LOOPBACK (Digital)

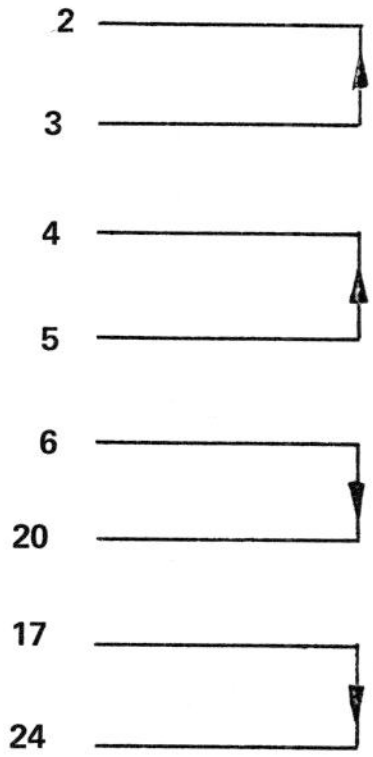

Modem strapped for external transmit clock

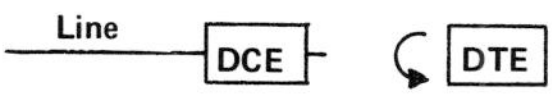

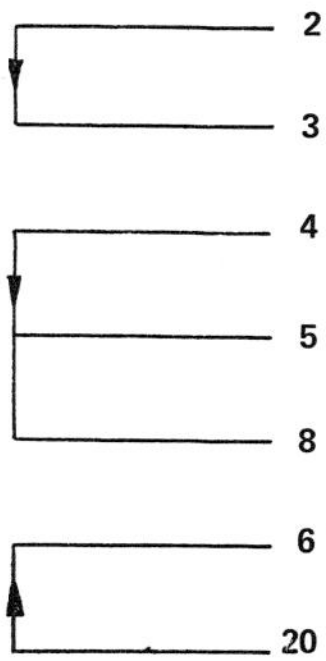

TERMINAL LOOPBACK

LOOPBACK TIMING

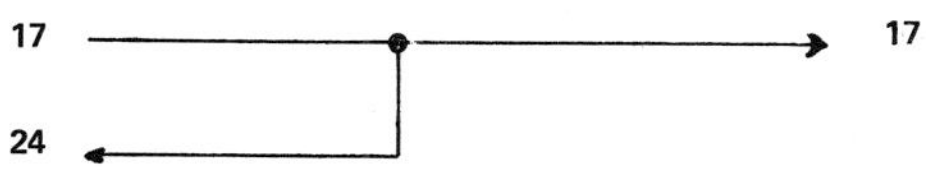

Other circuits as normal

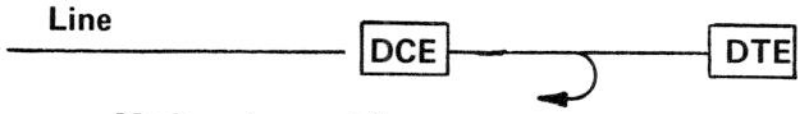

Modem strapped for external transmit clock

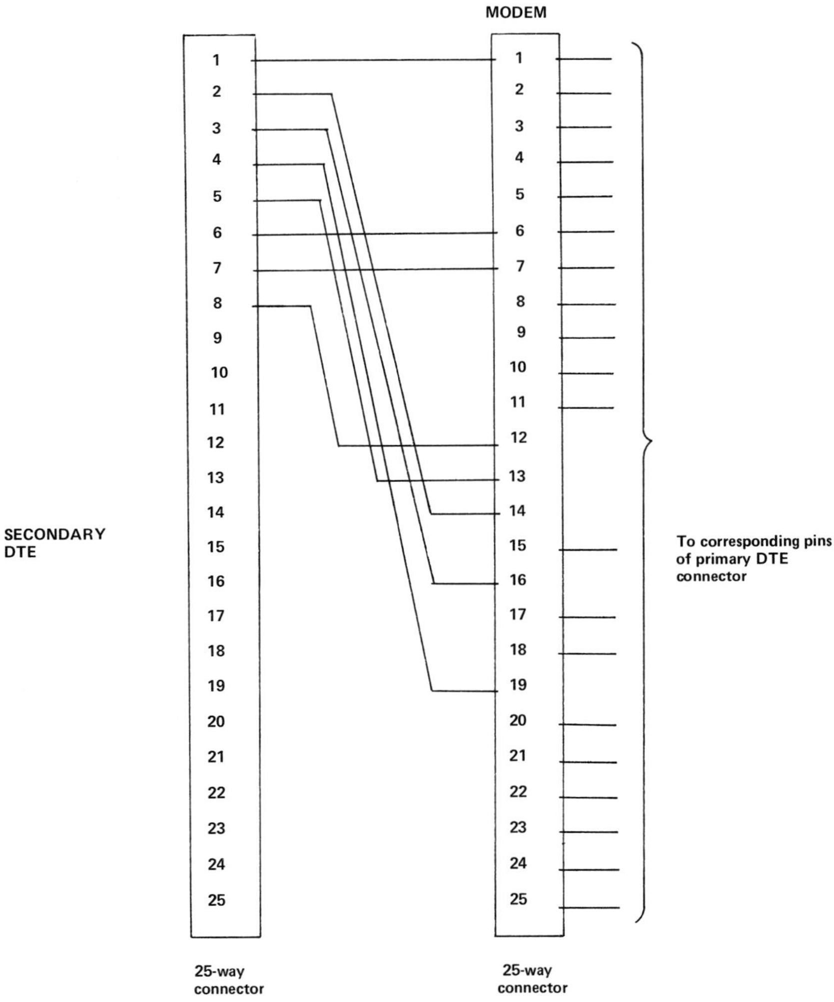

Connector Wiring for Secondary Forward Channel

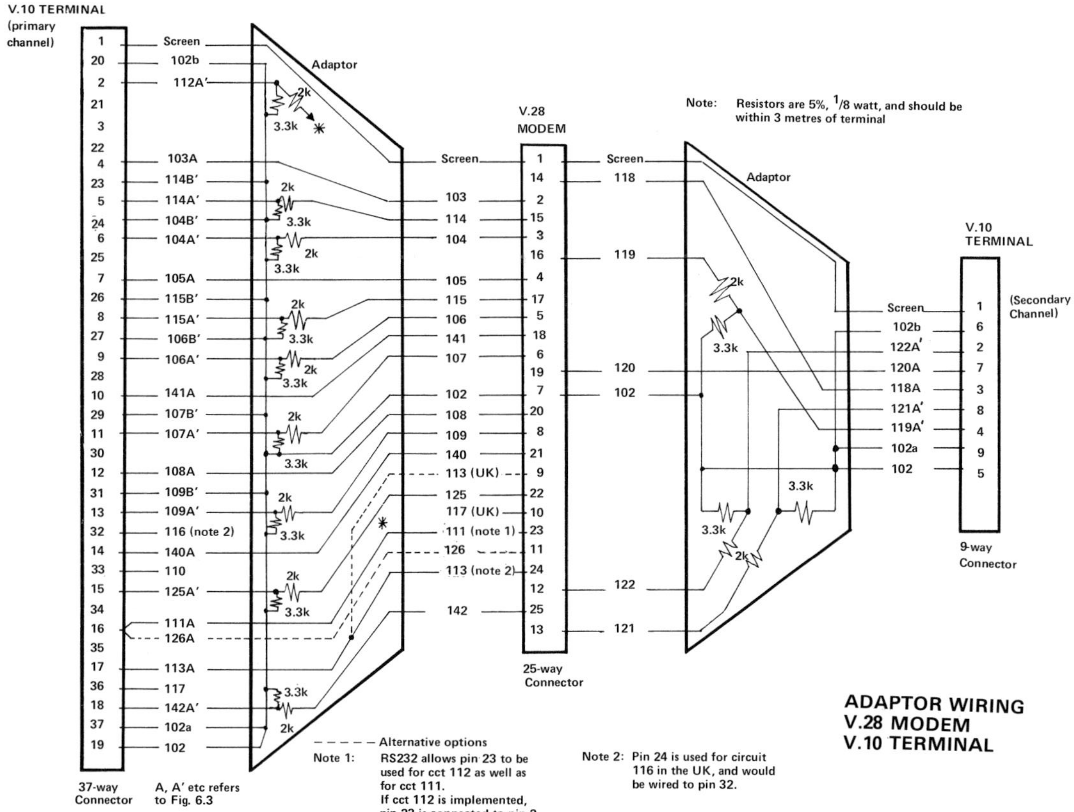

– – – – – Alternative options

Note 1: RS232 allows pin 23 to be used for cct 112 as well as for cct 111.
If cct 112 is implemented, pin 23 is connected to pin 2, not pin 16 as shown.

Note 2: Pin 24 is used for circuit 116 in the UK, and would be wired to pin 32.

Note: Resistors are 5%, 1/8 watt, and should be within 3 metres of terminal

ADAPTOR WIRING
V.28 MODEM
V.10 TERMINAL

Note: Resistors are 5% 1/8 watt, and should be within 3 metres of modem.

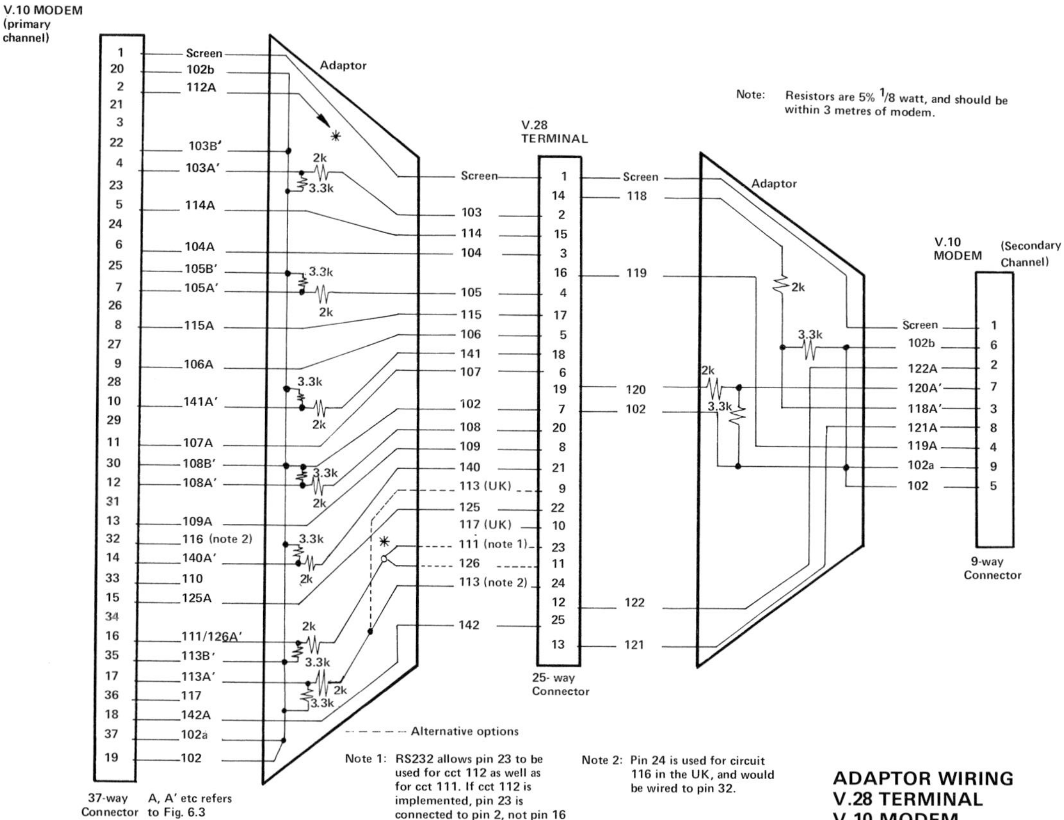

---- Alternative options

Note 1: RS232 allows pin 23 to be used for cct 112 as well as for cct 111. If cct 112 is implemented, pin 23 is connected to pin 2, not pin 16 as shown.

Note 2: Pin 24 is used for circuit 116 in the UK, and would be wired to pin 32.

ADAPTOR WIRING
V.28 TERMINAL
V.10 MODEM

Appendix 3

Comparison of Standards Terminology – CCITT REC V.24, EIA RS232C, EIA RS499

EIA RS232C		CCITT V.24			EIA RS499	
Signal ground	AB	102	Signal ground or common return	102	SG	Signal ground
		102a	DTE common return	102a	SC	Send common
		102b	DCE common return	102b	RC	Receive common
Transmitted data	BA	103	Transmitted data	103	SD	Send data
Received data	BB	104	Received data	104	RD	Receive data
Request to send	CA	105	Request to send	105	RS	Request to send
Clear to send	CB	106	Ready for sending	106	CS	Clear to send
Data set ready	CC	107	Data set ready	107	DM	Data mode
–		108/1	Connect data set to line	108/1	–	
Data terminal ready	CD	108/2	Data terminal ready	108/2	TR	Terminal ready
Received line signal detector	CF	109	Data channel received line signal detector	109	RR	Receiver ready
Signal quality detector	CG	110	Data signal quality detector	110	SQ	Signal quality
Data signal rate selector (DTE)	CH	111	Data signalling rate selector (DTE)	111	SR	Signalling rate selector
Data signal rate selector (DCE)	CI	112	Data signalling rate selector (DCE)	112	SI	Signalling rate indicator
Transmit signal element timing (DTE source)	DA	113	Transmitter signal element timing (DTE)	113	TT	Terminal timing
Transmission signal element timing (DCE source)	DB	114	Transmitter signal element timing (DCE)	114	ST	Send timing
Receiver signal element timing (DCE source)	DD	115	Receiver signal element timing (DCE)	115	RT	Receive timing
–		116	Select standby	116	SS	Select standby
–		117	Standby indicator	117	SB	Standby indicator
Secondary transmitted data	SBA	118	Transmitted backward channel data	118	SSD	Secondary send data
Secondary received data	SBB	119	Received backward channel data	119	SRD	Secondary receive data
Secondary request to send	SCA	120	Transmit backward channel line signal	120	SRS	Secondary request to send
Secondary clear to send	SCB	121	Backward channel ready	121	SCS	Secondary clear to send

EIA RS232C		CCITT V.24			EIA RS449	
Secondary received line signal detector	SCF	122	Backward channel received line signal detector	122	SRR	Secondary receiver ready
		123	Backward channel signal quality detector	123	–	
		124	Select frequency groups	124	–	
Ring indicator	CE	125	Calling indicator	125	IC	Incoming call
		126	Select transmit frequency	126		
		127	Select receive frequency	127		
		128	Receiver signal element timing (DTE)	128		
		129	Request to receive	129		
		130	Transmit backward tone	130		
		131	Received character timing	131		
		132	Return to non-data mode	132		
		133	Ready for receiving	133		
		134	Received data present	134		
		135		135	IS	Terminal in service
		136		136	NS	New signal
		140	Remote loopback for point-to-point circuit	140	RL	Remote loopback
		141	Local loopback	141	LL	Local loopback
		142	Test indicator	142	TM	Test mode

Index